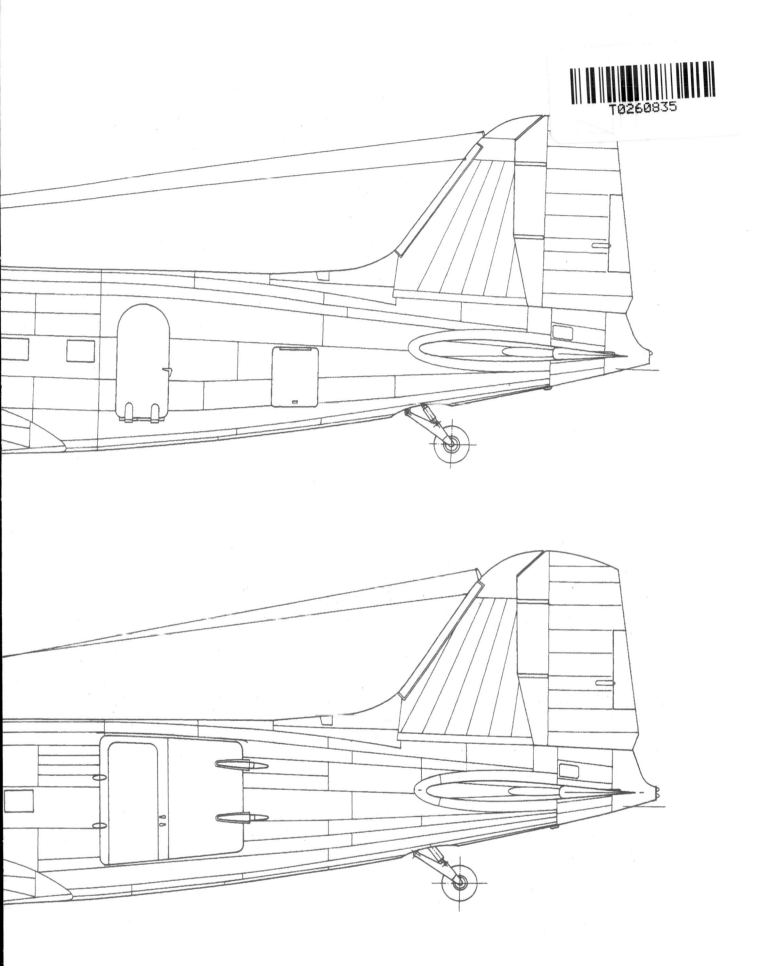

THE LEGENDARY
DOUGLAS DC-3

A PICTORIAL TRIBUTE

DAKOTA - GOONEY BIRD - SKYTRAIN - SPOOKY - SKYTROOPER

A PHOTOGRAPHIC TRIBUTE ACROSS THE GLOBE ON
THE ICONIC DOUGLAS DC-3 & C-47
'THE LEGEND GOES ON'

THE LEGENDARY
DOUGLAS DC-3
A PICTORIAL TRIBUTE

MICHAEL S. PROPHET

Lanasta

Lanasta

ISBN: 9789464560640
e-ISBN: 9789464560657

NUGI: 465
1st print run, November 2022

© Copyright 2022
Walburg Pers / Lanasta

DTP: Jantinus Mulder
Photos: Michael Prophet
Proof read: Tony Merton Jones

All patches & stickers: M. Prophet Collection

Category:
Historic Transport Plane - Douglas, Dakota,
DC-3/C47, Super DC-3, Piston twin engine
propliner - Vintage transport.

Center Photo:
Mid America Flight Museum N5106X, Mount
Pleasant Texas

Printed in the Netherlands.

All correspondence regarding copyrights, transla-
tion or any other matter can be directed to:
Walburg Pers / Lanasta,
Nieuwe Prinsengracht 89,
1018 VR Amsterdam

www.lanasta.com

FOREWORD

It gives me great pleasure to introduce a book written and compiled by one of the world's most talented and dedicated aviation photographers. For many years I have been impressed by Michael's intense passion for the Douglas DC-3, and in his quest to shoot the aircraft in as many locations around the world as possible he has become one of the leading authorities on the type. His countless contributions to "Propliner" over the years have helped to create a loyal readership all over the world and here is an opportunity to enjoy even more of his magical photographs.

Surely one of the world's most iconic aircraft, the Douglas DC-3 has captivated enthusiasts for generations. More than eighty years after her maiden flight, the type is still making history today, and the recent round-the-world flight by a 1939-vintage example is surely testimony to the robust nature of the aircraft's design and its enduring popularity.

Originally purchased by American Airlines as the Douglas Sleeper Transport, the airliner soon evolved into the DC-3 that we know and love today. By the time that war broke out in 1939, DC-3s were in service with many other American airlines including United, Eastern and Braniff, while across the Atlantic in Europe the DC-3 had instant appeal, with KLM, SABENA, Sweden's ABA, Swissair and others all acquiring the revolutionary new airliner. Of course, the war changed everything, and overnight the DC-3 found an important new role as military transport. Serving in large numbers with the Allies, the aircraft became known as the Dakota, C-47 'Skytrain' or 'Gooney Bird', and this versatile flying machine became the backbone of the transport force. It took part in many famous operations, the most notable of which was the Allied invasion of northern France, when on D-Day - June 6 1944 - a grand total of 821 Skytrains and Dakotas dropped 13,000 paratroops over enemy lines. There were many other occasions when the aircraft flew her way into the history books, but none was as poignant or memorable as the D-Day operation.

Post-war the DC-3 helped to restore many vital air services around the world, as the arrival of peace and the need to establish aerial communications between countries as quickly as possible ensured that the DC-3 was given this task until new airliners became available. Although many designs were marketed as a 'DC-3 replacement', it took several decades before the type was replaced in airline service, while air forces around the world continued to take full advantage of this rugged transport for many years. Indeed, the type was so popular that turboprop conversions still fly today with numerous air forces across the globe.

Everyone must have their own special DC-3 memory, and for me it was my first ever flight in a DC-3 nearly fifty years ago. It was a sunny August afternoon and during a stroll through the terminal at East Midlands Airport I came upon a rather scruffy individual in an old green sweater selling tickets for a pleasure flight aboard Kestrel Aviation's former BEA "Pionair" G AMFV. Seizing the opportunity, a ticket was promptly purchased and minutes later a small group of passengers was led out to the aircraft parked on the main apron alongside many more modern jet and turboprop types. Having taken our seats in the cabin, I was amazed to see the green jersey-attired ticket seller making his way up to the cockpit, where he duly took his place in the captain's seat and proceeded to convey his clients on the most enjoyable of pleasure flights ever. Our twenty-minute aerial excursion over the Derbyshire countryside preceded a landing in the wake of a British Midland BAC One-Eleven jet arriving from Jersey - a routine daily adventure for the DC-3 just a few years earlier.

As you thumb your way through this book please imagine the effort that has gone into the creation of each and every picture, and the pleasure that it has given the photographer.

Tony Merton Jones
Editor - Propliner
Salisburym January 2022

ABOUT THE AUTHOR

From a young age, Michael S. Prophet became fascinated with aviation, when he started drawing WW2 airplanes at primary school. Born on the Caribbean Island of Curacao, Dutch Antilles (1956), his first introduction off flying was when, as a baby his mother took him for a famlily visit to Trinidad, Tobago onboard a twin engine KLM Convair 440.

After moving to The Netherlands with his parents, his love for aviation really kicked off when he purchased his first transatlantic ticket with Pan Am, flying on a Boeing 707 jetliner visiting friends in New York in 1974. This also triggered the interest in taking photographs. Since then much of his spare time, has been devoted to aviation/travel and photography.

Just out of high school, he started his career at KLM Royal Dutch Airlines as a junior aircraft mechanic, working on Douglas DC-8 and B747/DC-10 wide-bodies Jumbo Jets. Later he started working at Dutch Fokker Aircraft Company, where he worked as an Interior Design Engineer on several new projects, such as the all new twin engine Fokker 50 and F100 jetliner.

In the early 1980s he bought a copy of the famous Osprey Colour Series 'Skytruck' book. The author Stephen Piercey, became like a mentor and inspiration to him. Admiring the amazing pictures he quickly developed his own style of photography. This was the moment that he started a life-long love for the dwindling number of classic jet & prop-liners around the globe.

Whilst on holiday in Hawaii, he photographed his first real operational Douglas DC-3 at Honolulu airport and a couple of years later, on the other side of the world, he got his first passenger ride on the vintage Douglas DC-3, onboard an Aero Virgin Island Airways flight from San Juan PR to St Thomas Virgin Islands ... and the rest is history as they say!

DEDICATION:

I am dedicating this book to my beloved mother *"Ena Yasmini Alexander"* (2018†) and younger brother *"Stuart Heston Schmidt"* (2020†), they both passed suddenly and way too soon. My mother's legacy sparked the interest in me for my global travels and photography.

website: Vintage Aviation Pictures - www.michaelprophet.com

Top:
The author at work, doing what he loves best, taking pictures of a DC-3. Hill Air Co, Fort Lauderdale Florida 1987

Photo: Paul van den Berg

CONTENTS

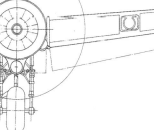

2ND DISCLAIMER

To the best of my knowledge the information in this book is true and complete. All data & information quoted are without any guarantee on the part of the author & publisher, who also disclaims any liability incurred in connection with use of this data or specific detail. We recognize that some words, model names and designations are the property of the trademark holder. This is used only for identification purposes.

BIBLIOGRAPHY/REFERENCE/SOURCE:

- *The Douglas DC-1/DC-2/DC-3, The First Seventy Years* Jennifer M. Gradidge Air Britain (2006) & *75 Years "Celebration Edition 3"* (2011)
- *The Legacy of the DC-3* - by Henry M Holden (1996) Wind Canyon Publishing Inc
- *DC-3 & C-47 Gooney Birds* by Michael O'Leary Motorbooks International (1992)
- *Skytruck* by Stephen Piercey Osprey Publishing Limited (1984)
- *Naval Fighter Series Douglas R4D-8/C-117D Super Gooney* by Steve Ginter (2013)
- *The Legendary DC-3* by Carroll V Glines and Wedell F Moseley - Bantam Air Series (1979)
- *Douglas - The Santa Monica Years* Marquand Books (2009)
- D-Day Squadron USA - Tunison Foundation
- Aviation Safety Network ASN
- Private files, travel notes & aircraft spotting logs
- Jan Koppen, Andre van Loon, Paul van den Berg plus many international aviation friends who provided input, technical support and relevant information.

DEVELOPMENT HISTORY

THE WINGS OF HISTORY

This magnificent aircraft has earned its place in aviation history, set many records and scored innumerable aviation firsts! It has flown more miles, piled up more flying time, carried more cargo and passengers then any other airplane in the world. It helped to lift America out of the depression and helped to make civilian Air Transportation to flourish across the globe. Its manufacture called it Douglas Commercial 3 (DC-3). It has many names such as: "Dizzy Three", "Old Methuselah", "Placid Plodder", "Dowager Duchess" and "Gooney Bird". The US Air Force called it "Skytrain" or "Skytrooper". The US Navy referred to it as a "R4D" and while in use as an AC-47 in Vietnam conflict it was referred to as "Spooky" or "Puff the Magic dragon". The British called it the "Dakota". During the Berlin Airlift she was referred to as "Rosinenbomber" as known as "Candy Bomber".

US PRODUCTION

It was clear that the Douglas Santa Monica plant, from which all DC-2s and early model DC-3s were manufactured, would not be able to meet the new delivery schedules and the increased production for US Government and later the WW2 effort. It was decided a new plant at Long Beach California was to be constructed. The first C-47-DL came off this line at the end of December 1941. A 3rd production line at Oklahoma City was set up and was used only for the production of C-47-DK "SKYTRAINS" and C-117-DK.

DO Santa Monica production,
DL Long Beach production
DK Oklahoma production

Typical civil designation: DC-3A
DC-3 = Douglas Commercial 3
A = production model/engine variant

Typical early model designation: DST-A-207A
DST = Douglas Sleeper Transport

Typical early model designation: DC-3-202A
202A = customer model

Typical military designation: C-47A-60-DL
C-47 = military version of the DC-3
A = production model
60 = USAF Block Number
DL = production plant

RUSSIAN PRODUCTION

The USSR bought 21 DC-3s direct from Douglas, through various trading companies between 1936 and 1939. At least two of these were delivered unassembled to serve as patterns for licence production. Production was initially undertaken at the State Aircraft Plant No 84 near Moscow and started in 1940. The first aircrafts were designated as PS-84 (Passazhirskii Samolet – Plant 84) and they were fitted with the 900 hp Shvetsov M-62 engines. Later in 1941 the production plant was forced to move to the Tashkent Plant no 18 and the aircraft designation was changed into Lisunov Li-2.

JAPANESE PRODUCTION

When the DC-2 was launched, Japan was looking for a replacement for their outdated Fokker Super Universal and Trimotor fleet. In March 1934 Nakajima Hikoki KK bought the licence right to build and sell the DC-2 in Japan. A single complete DC-2 was delivered during November 1934. Several DC-2s were delivered, but the DC-3 showed greater promise and that ended the DC-2 production. Mitsui purchased the production rights for the DC-3 in February

1938. The first aircraft assembled from American components emerged in September 1939 for Japan Air Transport Co. but the 2nd and 5th aircraft went to the Navy! The imported DC-3s were known as D1 and the Japanese production line labelled them D2…later L2D. The Allies used the nickname "Tabby". The L2D were fitted with 1000 hp Mitsubishi Kinsei 43 engines.

Production Figures (Air Britain DC-1, DC-2, DC-3 the First Seventy Years 2006)	
Santa Monica Civil DC-3	579
Santa Monica Military	382 (C-41, C-41A, C-53, C-53D and R4D-2)
Long Beach Military	4285 (C-47, C-47A, C-47B and R4D-1)
Oklahoma City Military	5381 (C-47A, C-47B, C-117A and R4Ds)
Oklahoma City Civil DC-3D	28
Grand Total	10,655
Japanese L2D	487
Russian Li-2	4937
World Total	16,079

Note: *During the hectic years of World War II, no less than 268 DC-3s (134 built at Long Beach and 134 at Oklahoma) were issued identical construction numbers (c/n). This error resulted in 134 pairs of DC-3s that had the same c/n, but were in fact different aircraft.*
This situation was officially corrected by Douglas in a Service Bulletin in July 1946. In this bulletin, Douglas determined that the c/n as allocated by the Long Beach plant were correct (and thus remained unchanged), and the c/n as allocated by the Oklahoma plant were revised. The corrected c/n are usually written as **13798/25243**, *i.e. incorrect number first, new number last.*

Page left
Spirit of Santa Monica Douglas C-53D Santa Monica Airport California

AIRFRAME MODIFICATIONS

Various modifications of existing airframes have been made either experimentally or for commercial use, by fitting non-standard engines and numerous other technical changes. This is a brief listing of major modifications.

Hi-Per DC-3

Modified by Pan American for use by Panagra and Avianca, with 1450 hp P&W R-2000-D5 engines & various airframe modifications to improve single-engined performance Also Known as a model DC 3-R2000

Mamba Dakota

Test-bed used for Armstrong Siddeley and fitted with 1425 hp Mamba ASMa.3 engines and later with the more powerful 1590 hp Mamba ASMa.6.

Dart-Dakota (1)

Test-bed modified by Rolls-Royce with 1540 hp Dart 504 engines and later with 1640 hp Dart 510 & 526 (one in each nacelle)

Dart-Dakota (2)

Modified by Field Aircraft Services for BEA trials before the Viscount deliveries and fitted with 1540 hp Dart 505s engines. (max weight: 28,200 lbs - 12,900 kg)

Abbreviations:	
c/n	construction number
s/n	serial number
a/c	aircraft
AF	Air Force
BuNo	Bureau Number
Aka	Also known as
STC	Structural Type Certificate
CAA	Civil Aviation Authority
CAA	Chinese Aviation Authority
Max	Maximum
MK	Mark
Hp	Horsepower
WW2	World War 2
RAF	Royal Air Force
LLC	Limited Liability Company
LH	Left Hand
RH	Right Hand
USN	United States Navy
USMC	United States Marine Corps
Ltd	Limited
Corp	Corporation
VIP	Very Important Person
USAAF	United States Army Air Force

Dart-R4D-8

Rolls-Royce Darts were fitted to N156WC for Pilgrim Airlines

USAC DC-3 Turbo Express

Aircraft fitted with PT-6A-45R engines and various airframe modifications such as square-tips tail-plane, new generators, electric, fuel, hydraulic and fire protection changes.

Basler BT-67

When Basler took over the Turbo Express DC-3 additional modifications were made. New 1424 hp PT-6A-67R engines were fitted giving the aircraft a payload of 13,000 lb. It was marketed as Basler Turbo-67 (BT-67) with a new STC February 1990.

Airtech

Modification consisted of replacing the R-1830s with WSK ASz-621Rs a variant of the Wright Cyclone R-1820 radial engine

AMI DC-3-65TP

Schafer Aircraft Modification Inc in conjunction with Aero Mod International (AMI) developed a turboprop conversion using the Pratt & Whitney Canada PT6A - 65AR engines. It was marketed as Schafer DC-3-65TP Cargo-master (1985). Kansas based Dodson International took over the AMI Supplemental Type Certificate (STC) and started DC-3 conversion (1997). Then in 2016 the

AMI STC was re-sold to Preferred Airparts LLC based at Kidron (OH) and remarketing it as the Preferred Turbine -3.

BSAS DC-3/C-47-65ARTP Aircraft

Turboprop conversion by BSAS International based at Wonderboom Airport South Africa. Based on a Wonder Air PTY Ltd modification STC for civilian and military use

Tri-Turbo-Three

The Tri-Turbo-Three (N4700C/N23SA) was a one-off conversion which reappeared in 1977. It's the only triple engined DC-3 which was fitted with the Pratt & Whitney PT-6A engines. The concept by Jack Conroy and Clay Lacy was first flown in 1977. It was marketed by Specialised Aircraft, Camarillo CA but never received a FAA certificate.

TS-62 = a hybrid soviet C-47B conversion

which included the ASh-62IR radial engines and extra LH side cockpit window. They were used by Aeroflot until 1957. Some were sold to china.

Canadair

Along with a large Douglas equipment stock, Canadair also purchased a large number of former USAAF C-47 Skytrains, as well as ex-Royal Air Force (RAF) Dakota transports. From 1945 to 1947, Canadair converted, modified, overhauled, and re-built several hundreds of C-47/DC-3 conversions for numerous domestic and international air carriers.

Scottish Aviation limited

The Scottish Aviation Dakota conversions where authorised by the Douglas Repair, Overhaul and Conversion centre for Great Britain. They could convert any DC-3 into four categories: regular freighters, standard 21 seater airlines configuration and Deluxe 21 seater and a VIP executive model.

Tropicana DC-3 (1957)

The Tropicana Conversion was designed, engineered and produced by L.B. Smith Air Corp based at Miami. This conversion consisted of replacing the forward cabin bulkhead with plexiglas panels. This created extra space and thus allowing extra forward windows. Extra side mounted divans and swivel chairs were fitted. In addition the more powerful Wright engines were installed increasing the cruising speed to 210 miles per hour.

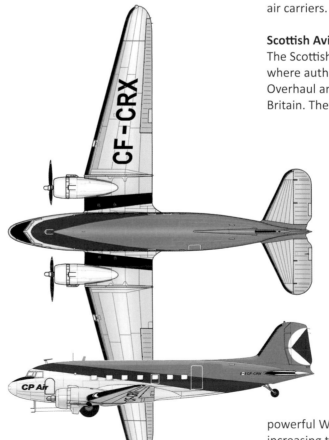

Artwork Andre Plourde

DOUGLAS AIRCRAFT COMPANY, SANTA MONICA CALIFORNIA, THE BIRTHPLACE OF THE DC-3

Douglas C-53-DO N242SM (c/n 4877) was originally delivered as a C-53 "Skytrooper" but converted to a US Navy R4D-3 with BuNo 05075 February 1942. Its initial role was a 28-seat paratrooper and glider tug aircraft. Retired from military service she was purchased by Nationwide Airlines with registration NC1075M August 1946.

This aircraft was been refitted with a AiResearch Maximizer kit, which included landing gear doors, short engine exhaust stacks, special engine cowling, oil cooler fairings and a large dome prop spinner. According to the AiResearch brochure this kit added a 20 mph speed increase and 200 miles range.

Painted in bold blue & red colors with Douglas *"Spirit of Santa Monica"* title's she is displayed at the Santa Monica Airport, CA and pays tribute to the company's founder *"Donald Wills Douglas"* and the historic nature of this airport. The first flight of a DC-3 took place on December 17, 1935 at the Santa Monica Airport. These aircraft were so successful and reliable that count-less DC-3 aircraft followed in production and remain in service throughout the world today. May 2017

Note: *The city of Santa Monica and the Douglas Aircraft Company are tied together in aviation history. Over 10,724 airplanes were produced ranging from the 1921 Cloudster to the B-18 Bolo, A-29 Havoc, DC-1, DC-2, DC-3, DC-4, DC-6 and the last piston airliner DC-7. The DC-3 monument is located at 3100 Airport Ave on the South side of the airport. Surrounding the aircraft are landscaped walkways, benches and a bronze sculpture of Donald Wills Douglas and his dog Bar.*

DOUGLAS DC-2

Only two DC-2s remain airworthy today, one of them is the **Aviodrome Museum** DC-2-142 NC39165/PH-AJU aka "Uiver" (c/n 1404). She was originally delivered to the USN Anacostia as a R2D-1 with BuNo 9993 delivered Sept 1935. It went to NAS Pensacola, August, 1940, and later to NAS Anacostia on January, 1941, before being struck off charge on August 28, 1944.

In April 1945 it was registered to its first civil owner as NC39165, during 1969 it was painted in General Air Lines colours and by 1983 it was owned by Colgate W. Darden III of Cayce, South Carolina, USA.

In November 1983 it was shipped to the Netherlands and was painted in a "Royal Dutch Air Lines" livery as PH-AJC "Uiver" with race number 44 for a special commemorative flight to and from Australia for the anniversary of the London to Australia air race of 1934.

She is now part of the Aviodrome Museum collection based at Lelystad airport Holland. She was the star of the Hamburg Airport Open day's airshow. September 2007

Page left
TransNorthern Aviation Douglas DC-3S Palmer Airport Alaska

Top: Lelystad Holland. Middle: Duxford UK Bottom: Zoersel Belgium

DOUGLAS DC-3 / C-47

EUROPE

DDA Classic Airlines flagship Dakota PH-PBA (c/n 19434) rolled off the Longbeach production line during January 1944 as a C-47A-75-DL "Skytrain" and was delivered to the USAAF with serial number 42-100971. She was transferred to the 8th/9th AF 44th TCS, 316 Troop Carrier Group (TCG) which was based in Cottesmore (Station 489) England. She took part in Operation Mission Boston, dropping 18 paratroopers in the early morning of June 6th 1944 over Ste Mere Eglise, 5 hours before the main invasion of Normandy. Several months later she participates in Operation Market Garden (17 Sept 1944) dropping paratroopers near Arnhem). She was named "The Squirrel" by her crew and she operated several supply missions from the provisional airfield at Keent near Grave the Netherlands.

After WW2 she served with the Dutch Royal Family as an executive transport as PH-PBA. It ended its flying day in 1975 when it was purchased by the Aviodome Museum at Schiphol Airport and put on static display with fake PH-TCB registration.

After spending 21 years on exhibition it was restored back to airworthy condition by Air Atlantique in Coventry UK. Parts of DC-3 G-DVOL wings & engine mounts were used to restore PH-PBA.

In 2006 PH-PBA was operated in a dual KLM/Air France retro livery to commemorate the first flight of a KLM DC-3 service to Paris. On the 7th December 2010 she was officially renamed "Princes Amalia" and repainted in the beautiful 1960's KLM – Royal Dutch Airlines colors.
Due to the loss of financial funding from KLM, DDA was not permitted to have KLM colors on the aircraft, so DDA decided (January 2018) to repaint the PBA back into her former 'Government' livery including the coat of arms of the Inspector General of the Dutch Armed Forces. As of April 2021, the DDA Classic Airlines relocated from Lelystad Airport back to Amsterdam Schiphol Airport. All technical staff, offices and DC-3 now reside inside the ex Martinair Hangar 32, situated at the General Aviation ramp.

DDA **Dutch Dakota Association**
First European Classic Airline

Martin's Air Charter (MAC) beautiful C-47A-80-DL PH-DDZ (c/n 19754) rolled of the Long Beach production line back in March 1944 with serial number 43-15288. She did not go to war and remained in the US with numerous squadrons.

After being stored at Davis Monthan AFB, Arizona for a short period, her civilian life started with the Federal Aviation Administration (FAA) as N161. In 1964 she went overseas to Somali Airlines with registration 6OS-AAA & 6O-SAA. It went to Egypt with Pyramid Airlines as SU-BFY and was known as "Khephren". In 1985 she returned to Malta Int'l Aviation Company (MIACO) and was offered for sale.

The Dutch Dakota Association (DDA) was looking for a long term DC-3 restoration project and purchased SU-BFY on February 13th 1987 and assigned the Dutch registration PH-DDZ. She was re-named "Sleeping Beauty" and arrived in Holland on May 10th the same year.

May 7th 1999 was a historic day; after a 12-year restoration PH-DDZ made its maiden test flight.

A Boeing (McDonnell Douglas) representative remarked that this "Grand Old Lady", as being the last built DC-3. This was a big complement to the DDA team.

The DDA painted DDZ retro Martin Air colors to represent Martin Air Dutch aviation history. She spent many years on the airshows circuit on display and giving rides from Schiphol Amsterdam Airport. She was sold to the Aviodrome Museum at Lelystad Airport in August 2016 and moved by a canal barge to its current location for further restoration in the museum. March 2012

2022 will be a festive year, the DDA Classic Airlines will celebrate its 40th anniversary. Back in March 1982 two pilots of the Dutch charter airline Transavia shared an idea to bring an airworthy Douglas DC-3 to the Netherlands and formed the Dutch Dakota Association (DDA).

Dutch Dakota Association
First European Classic Airline

DDA CLASSIC AIRLINES
25 YEARS

Two logos of 2007.

Page left:
Top: Caen Carpiquet, Normandy France

Bottom:
Cherbourg – Maupertus France

The French Association **"Un Dakota sur la Normandie"** C-47B-35-DK F-AZOX (c/n 16604/33352) rolled off the Oklahoma production line with s/n 44-77020 in May 1945.

She was assigned to RAF Montreal as KN655 and later transferred to the Canadian Air Force with tail code 1000/12965. Her military career ended in October 1974 and she entered civilian life with Eclipse Consultants Ltd as C-GGJH based at Oshawa Canada. In September 1979 she was converted from a DC-3A to a DC3-S1C3G by Skycraft Air Transport Canada. They re-installed a LH passenger door and converted it back into a passenger configuration. Air Dakota bought her in 1991 and flew her to France as F-GIDK.

In 2009 she was bought by Mr Alain Battisti (president of Chalair) and moved her to Melun Villaroche Aerodrome Southwest of Paris. F-AZOX is privately owned and used for airshow, film, aviation promotional use. September 2016

Top: Antwerp Belgium

Bottom: Duxford IWM UK

To mark the 2015 anniversary of Operation "Market Garden" **Aces High** Skytrain C-47A-75-DL N147DC (c/n 19347) aka "Mayfly" visited Holland and took part in the commemoration parachute drop flights organized by the RCPT from Kempen-budel Airfield Holland.

She rolled of the Longbeach production line for the USAAF with s/n 42-100884. (Dec 1943) and joined the 8th/9th AF – 438 Troop Carrier Group in 1944. Later she operated with the RAF as TS423 based at Netheravon England and shortly before the end of the war she was transferred to 436 Squadron to the Royal Canadian Air Force at Down Ampney.

Center: Duxford UK

After WW2 she took part in the Berlin Airlift operation. She stayed in England and served several companies: Short Brothers, Scottish Aviation, Marshalls and Ferranti as G-DAKS. At Ferranti she acquired a non-standard nose which housed large air pass radar.

Mike Woodley (Aces High Ltd) bought it and used her in several TV series such as "Vera Lynn" of Ruskin Air Services in the series AIRLINE in 1979. She was featured in the international movie Band of Brother in USAAF colours. She remains a young lady with just over 3,500 hours total flying time, probably the lowest hour Dakota flying in the world.

In 2019 she was painted back in her own original D-Day colours with code S6-A and Ace poker card logo to commemorate the 75th Anniversary of D-Day.

Top: Duxford UK

Danish Dakota Friends (**DC-3 Vennerne**) OY-BPB (c/n 20019) was built at Douglas Long Beach, California as a C-47A-85-DL Skytrain and was accepted by the USAAF with serial number 43-15553 (April 1944).

At the end of the war she flew with the Royal Norwegian Air Force (RNAF) with tail code 315553/X-20 until 1946 and then with DNL-Det Norske Luftfartsselskap as LN-IAT. In 1953 she flew with the Danish Air Force with s/n 68-682 later changed to K-682.

During the mid 1980s she was leased by Bohnstedt-Petersen A/S but retained the RDAF colors. Between May 1996 & August 1997 she was repainted in a SAS Scandinavian Airline System livery and carried the name "Arv Viking" on the nose.
She is a regular airshow visitor and as such was one of the partitipants at the 2014 "Daks over Normandy" event at Cherbourg France. In addition she also joined the noteworthy 2019 "70th anniversary of D-Day" held at Duxford (UK) & Caen Normandy (France).

During a family holiday to the Spanish Island of Majorca I was able to photograph the "FLY LPI" C-47B-35-DK ES-AKE (c/n 16697/33445) aka "Congo Queen", at the local San Bonnet airport. This aircraft belonged to the **Vallentuna Aviation Association** based at Vallentuna airfield near Stockholm Sweden and carried the registration ES-AKE which later changed to 9Q-CUK.

During 2016 she received a new US registration N41CQ. On a flight to the UK she made a fuel stop at Lelystad Airport, Holland and this is where the author photographed her. Sept 2012

(ps: The Swedish owner sold her to a private owner in China and ferried her to Beijing during the fall of 2019. She now carries a period China National Aviation Corporation livery and a fake XT-125 registration)

Engine trouble brought an unexpected visit of the **Swedish Flygande Veteraner SAS** DC-3 "Fridtjof Viking" SE-CFP (c/n 13883) to Lelystad Airport Holland. They were on a VFR flight to England and diverted to Lelystad and became a welcome guest at the DDA Classic Airlines, for an engine change.

SE-CFP is a C-47A-DL Skytrain delivered to the USAAF with serial number 43-30732. Daisy was shipped directly overseas in October 1943 and was based in Oran in Algeria. She was involved in transporting troops, cargo and supplies missions. It joined the 9th Air Force, Feb 1944 and later the 53 Troop carrier Squadron at the end of 1945. She participated in Mission Boston on Tuesday, June 6, 1944 (D-Day) with an air landing of troops at night north of Amfreville in Normandy. In addition she also took part in operation Market Garden, 17 September that same year.

She returned to the United States in September 1945 and was sold to Canadair Ltd, which in 1946 modified it to the DC-3C. After the war it flew with the Norwegian operator Det Norske Luftfartselskap AS as LN-IAF based in Norway. It was then transferred to the Scandinavian Airlines System (SAS) and leased to Linjeflyg as SE-CFP. In 1960 it was sold to the Royal Swedish Air Force as 79006 and based with F7 Wing.

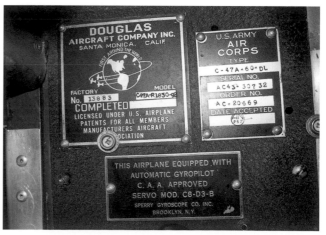

Air Atlantique Skytrain C-47B-15-DK G-AMPY (c/n 15124/26569) was delivered to the USAAF with s/n 43-49308 and transferred to RAF Nassau as KK116 late November 1944. After the war she stayed in the UK and served a number of companies such as Starways Ltd, Aviation Overhauls and Intra Jersey. Early 1980s she joined Air Atlantique and she was painted in period "Northwest" colours promoting the US Northwest Airlines.

Later she returned into the company's green and white Pollution Control colours and then it was repainted in its current RAF Transport Command scheme. At present she is part of the RVL Aviation Ltd fleet and on standby for pollution duties at Coventry. Her parent company has since moved on to B727 jets so it's unsure what the future holds for G-AMPY. During 2020 she was sold to a private owner in the USA, but due to the global Covid-19 crisis she remained stored at Coventry Airport UK.

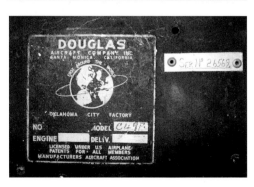

Update:
May 2022; G-AMPY was ferried to its new owner in Arlington WA (USA)

Top: Hamburg 2007

Left: Brussel 1987

France DC-3 Association F-AZTE was built as a C-47A-1-DL Skytrain (c/n 9172). It was delivered to the USAAF with serial number 42-23310 in March 1943 and went to the 8th Airforce during November 1943. After WW2 she flew with Scottish Airlines as G-AGZF and later she was used by the French Airforce as WZ984. During the 1980s she flew with Hemet Exploration (F-ODQE) and ACE Transvalair as F-GDPP.

She is currently based at Paris-Orly airport and flies in Air France colors with large F-BBBE registration on its fuselage. In 2007 she visited the Aviodrome Museum at Lelystad Holland painted in a dual KLM – The Flying Dutchman and Air France colors, for the Giant of History Fly In.

She was one of the honorary guests at the 70th anniversary of D-Day landing at Cherbourg airport Normandy. June 2014

Ex **Air Atlantique** DC-3 G-ANAF (c/n 16688/33436) aka "Pegasus", was photographed with her bold black and red livery after she was converted into a surveillance/pollution control sprayer with the RVL group. Coventry UK 2014.

Built as a C-47B-35-DK Skytrain with USAAF s/n 44-77104 she was transferred to RAF Montreal with tail code KP220 (June 1945). After WW2 she stayed in the UK and was operated by BKS Aerocharter Ltd with civil registration G-ANAF. She joined Air Atlantique Limited in 1977 and carried several different liveries' including Pollution Control and RACAL title's. An American company called Starflite Corporation showed interest in her and re-registered it N170GP, but the sale fell through and she remained in the UK.

Early 2019 she was purchased by the UK "Aero Legends" and joined their fleet of historic WW2 airplanes. She was repainted in olive drab D-day colors with her original WW2 RAF KP220 markings and flew over to Normandy for "75th anniversary" of D-Day (June 2019).

Bottom: Caen, Normandy

Top: Salzburg Austria 2014

GOLDTIMER
FOUNDATION

Currently only one Lisunov Li-2 remains in airworthy condition and it belongs to the **Goldtimer Foundation** based in Budapest Hungary. Painted in vintage Malev colours, Li-2T HA-LIX (c/n 184 332 09) was delivered by the Moscow Khimki plant 84 during September 1949. She joined the Hungarian Air Force with tail-code 209.

She regular participates in air show's and pleasure flights across Europe. She participated in the 70th D-Day Anniversary at Cherbourg Normandy, back in June 2014 and recently in the 75th D-Day Anniversary at Duxford UK and Caen Normandy France. June 2019

The immaculate **Breitling** DC-3-277B HB-IRJ (c/n 2204) first flew for American Airlines as NC25658 "Flagship Cleveland" back in March 1940. During the war years she was enlisted by the USAF and was used as an Army Personal Transport. After WW2 she was converted to a DC-3A, but she retained her standard right-hand side passenger door. Her civil career started with Trans Texas Airways and Provincetown-Boston Airlines (PBA) with registration N34PB.

As such she migrated to Florida, mostly during the winter period with Eastern Express/Bar Harbor Express. In July 1992 after a period of storage she was once again taken back in service Champlain Air, with a new registration N922CA aka "Priscilla". In 2008 she was purchased by Francisco Agullo and a group of friends and was ferried it

to Opa Locka Florida, where she received a major overhaul and fresh new colors for a new career in Europe. In 2017 she embarked in a World Tour which lasted six months and covered a distance of 45,400 km's (24.500 nm), crossing 27 countries and visiting 62 cities. After this remarkable achievement she was sold to the "MSO Air and Space Museum" based in Sivrihisar Turkey and is regular flown at airshow's.

Page Left – Bottom RH photo:
The gleaming Breitling DC-3 HB-IRJ visited the Zoersel Fly-In show, in Belgium back in the summer of 2012. I managed to capture the friendly brunet model Miss Laura Antonacci, fully outfitted in period 1950s uniform, on-board this VIP executive transport.

Flight Certificate
We certify that
MICHAEL PROHPET
Flew aboard the Breitling DC-3
Flight: LOCAL FLIGHT DOLE
Your Crew
Captain　　First Officer　　Cabin Attendant
Date: 28 SEPT 15

The Breitling DC-3 HB-IRJ took part in the 2014 "Daks over Normandy" D-Day event, from Cherbourg -Maupertus Airport France. I was invited by Captain Francisco Agullo to participate in a local scenic flight over the Normandy beaches as a tribute to the US 82nd Airborne and 101st Airborne Divisions C-47 paratroopers that jumped on the morning of the 6th June 1944.
Our route took us overhead the well-known town of Sainte-Mère-Église and Utah Beach. Captain Francisco Agullo was joined by Captain Jon Corley and flight attendant Rebecca Williamson while in the cabin some friends of mine Paul van den Berg, Roger Syratt and Djorn Kannengiesser enjoyed some cool champagne during this historic flight.

In 2015 the Breitling DC-3 celebrated the Douglas DC-3, 80th Anniversary of its first flight which took place on the 17th December 1935. A special souvenirs book called "DC-3 backstage" was published by owner/

pilot Francisco Agullo and it officially went for sale at Dole Airport France.

The Breitling DC-3 visited the 2016 addition of the Airliner Classics at the Speyer Airport, Germany. German born Stella Heidorn who is a big DC-3 fan, visited Speyer as she was reporting this event for her aviation blog. With thanks to the owner/captain Francisco Agullo, I invited Stella on-boards for an spontaneous photoshoot.

Page left, top: Dole, France 2015

Center: Zoersel, Belgium 2012

Left: Dole, France 2015

This beautifully restored sparkly DC-3 N431HM ex HB-ISC (c/n 9995) painted in Swiss Airlines colours was photographed at Salzburg W.A. Mozart Airport Austria.

Delivered as a C-47A-45-DL for the USAAF in August 1943 with s/n 42-24133 she was transferred to 85th Troop Command Squadron (TCS). Sometime between 1945 and 1948 she was converted to a passenger DC3C-S1C3G which included panoramic windows and a LH passenger door.

In 1986 she was bought by **Classic Air** (ex N88YA) who ferried it to Switzerland for a total overhaul. During the 1980s she regu-larly visited European airshows and perfor-med local pleasure flights. From 2004-2006 it was operated on a commercial basis by Ju-Air Switzerland.

She is presently privately owned by Hugo Mathys and was re-registered as N431HM. In 2010 she was ferried to Basler Turbo Conversions, Oshkosh USA and received a second major overhaul. This included the removal of the panorama windows and the installation of a new cockpit.

During June 2019 she attended the 75th Anniversary of D-Day, at both locations at Duxford (UK) and Caen Normandy, together with 27 sister ships.

Dakota Norway shiny bare metal LN-WND (c/n 11750) aka "Little Egypt" was built as a original C-53D-DO Skytrooper and delivered to USAAF 8th Air Force in June 1943 with s/n 42-68823. She later served with Finnair as OH-LCG and the Finnish Air Force as DO-9.

Norwegian pilots Thore Virik and Arne Karlsen, who also founded the Foundation and a member's organization in 1986. Dakota Norway is a non-commercial organization.

She took part in both events flying at Duxford Airfield (UK) and Caen Carpiquet Airport Normandy France, celebrating the "75th Anniversary" of D-Day June 2019.

Aero legends C-47A-75-DL N473DC (c/n 19345) Skytrain was part of the "**Dakota Heritage Inc**" Trustee based at Lincolnshire Aviation Heritage Centre, who initially brought her to England.

Delivered to the USAAF in December 1943 with s/n 42-100882 she was assigned to the 8th AF Feb 1944 and later to the 9th AF Troop Carrier Group and flew to the England in February 1944.
During Operation Elmira, 6th June D-Day, she carried soldiers of the 82nd Airborne Division on re-supply mission's and in addition also recovered WACO Assault gliders from Normandy landing grounds.

During September 1944 she was transferred to the RAF and redesigned as a Dakota Mk.3 with serial number TS422. During Operation Market-Garden she was used as a re-supply and gliders recovery transport.

After the war she operated for many years in Canada flying for numerous operators as a passenger airliner and freighter as C-FKAZ. In 1967 she was converted to a DC-3C-SIC3G by Northwest industries LTD Edmonton Canada. In 1985 she entered US registry as N5831B and was operated by McNeely Charter Service Inc.

She is currently painted in olive-drab D-Day colours and known as "Drag-em-Oot". Her last major appearance was at Caen-Carpiquet Airport Normandy France, celebrating the 75th anniversary of D-Day last year June 2019.

Top: Eindhoven, Netherlands 2014

Airveteran highly polished OH-LCH rolled of the Santa Monica production line, CA as a DC-3A-453 with (c/n 6346). Intended for Pan American Grace Airways as NC34953, she was promptly assigned to military service, re-designated as a C-53C with s/n 43-2033 (December 1942).

She served in the US Air Transport Command, North Atlantic Wing and was probably based at Presque Isle, Maine until transferred in November 1943 to the European Wing ATC, where she is believed to have served in personnel transport duties. In October 1944 she was transferred to the 8th AAF until the end of war in Europe and went into storage at Oberpfaffenhofen, Germany.

Aero OY/Finnair became the new owner and as OH-LCH and she made her first commercial flight for Aero Oy on 21st June 1948. She logged a total of 22137 hours until dismantled for spare parts on 15 December 1960, but was later rebuilt as a freighter with a large cargo door.

On 1st April 1967 she made Finnair's last scheduled DC-3 passenger flight. Having logged 28826 hours she was sold to the Finnish Air Force on March 1970 and received call sign DO-11. The Air Force retired all DC-3s in 1985 and DO-11 was sold to Airveteran Oy on January 1986 to become OH-LCH once again.

She regularly visits European airshow's and as such she took part in the 2006 Aviodrome DC-3 Fly-In (Holland), the 2014 Daks over Normandy (Cherbourg Normandy) and the 2019 75th Anniversary of D-Day at Duxford (UK) and Caen-Normandy (France).

Bellow: Caen, Normandy 2014

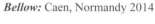

The **Battle of Britain Memorial Flight** (BBMF), C-47A-60-DL "Skytrain" ZA947 (c/n 10200) was photographed during an exhilarating take-off from Teuge Airport Holland. She visited Holland for the Dutch Queen Birthday and National holiday.

Originally delivered to the USAAF with s/n 42-24338, September 1943, she was assigned to the Royal Canadian Airforce 3rd Troop Carrier as 661/KG661 that same year. As a Dakota III she mainly served in Canada until 1967.

The Royal Aircraft Establishment (RAE) based at Farnborough then purchased the aircraft, which was allocated the UK military serial number KG661 and later changed to ZA947 November 1987. During the early 1980s she was painted in a white & grew RAE livery. Then by the mid 1980 she received a bright red and dark blue 'raspberry ripple' Royal Aircraft Establishment colors.

Currently ZA947 is painted in WW2 D-Day colors and is named 'Kwicherbichen'. She is displayed on her own and in combination with other members of the BBMF fleet. She is mainly used for training pilots and aircrew for the BBMF flagship the "Avro Lancaster".

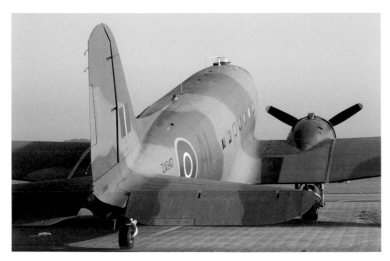

D-DAY SQUADRON

USA

The D-Day Squadron is part of the Tunison Foundation, a non-profit 501(c)(3) charitable organization. In June 2019, the D-Day Squadron led an American fleet of 15 restored Douglas C-47 and DC-3 variants, to take part in a flyover with more than 30 international aircraft, to drop more than 200 paratroopers over the original 1944 drop zones in Normandy, commemorating the 75th anniversary of D-Day. In addition it also celebrated the 75th anniversary of the Berlin Airlift at sites throughout the UK, France, Germany and Italy.

The Squadron's 'living history' program takes the compelling story of the citizen soldier to audiences at airshows and events off the flight line, to honor these brave Americans and ensure their memory and significance is appreciated for future generations. The group's efforts are funded through the generous tax-deductible contribution of their supporters. The Squadron represents the unity of flying vintage DC-3s, their crews and the opportunity to use these flying museums as an educational platform involving youth all over the world. Learn more at **DDaySquadron.org.**

Bottom: Weeze, Germany 2017

Dynamic Aviation Inc spotless C-47A-DL Skytrain N47E (c/n 13816) also known as "Miss Virginia" was originally delivered by Douglas Long beach, California in September 1943 and she went to the US Army AF with s/n 43-30665 and was assigned to ATC Morrison June 1944. She spent her military career in the USA and only ventured overseas to South America.

During her civilian life she was converted to a DC3C-S4C4G by Jungle Aviation & Radio Service Waxhaw, NC December 1978. Later on she received a civil registration N48065 until sold to Colombia as HK-2540P with "Instituto Linguistico de Verano" July 1981. She came back to the US late 1988 with JAARS Inc and K&K Aircraft Inc as N7043N. She belongs to Dynamic Aviation based in Bridgewater Virginia. It was restored to current beautiful USAF post-war color scheme with polished skin.

As part of the D-Day Squadron's Mighty Fifteen DC-3/C-47 fleet, she participated in the North Atlantic crossing also known as the historic WW2 "Blue Spruce" route to bases in England. The American contingent plus other European sister ships took part in the leading event of the year, which was the "75th Anniversary of D-Day", which took part at Duxford Airfield UK and Caen Carpiquet Airport Normandy France.

Betsy's Biscuit Bomber N47SJ C-47B-5-DK (c/n 14424/25869) rolled off Douglas Aircraft's production line in Oklahoma City, Oklahoma during the summer of 1944. The U.S. Army Air Force accepted her on September 4th, 1944 with s/n 43-48608. She served with the ATC 9th Air Force in Europe. Her nickname, "Betsy's Biscuit Bomber", derives from her time when she took part in the Berlin Air Lift during 1948.

After the war she joined the Belgium Airforce as K-11 and operated her with OT-CWF markings. Then she joined the French Air Force with s/n 348608 May 1953.

10 years later she joined the Israeli Defense Force with s/n 1416 with additional tail marking 4X-FNN. After military service (1993) she was stored at Lod Air Force base and in Sept 1999 sold to Global Aircraft Industries of Arizona USA received a new registration N47SJ. She was ferried to Villeneuve Alberta Canada in anticipation of onwards sale.

Bottom: Caen, Normandy 2019

Gooney Bird Group Inc, Paso Robles, CA purchased her back in 2007 and ferried it to California. In 2009 she was converted to DC3C-R-1830-90C specification. She has been a military girl all her working life and, as such, is one of the most original C-47s left – as well as being one of the lowest time airframes.

As part of the D-Day Squadron's Mighty Fifteen DC-3/C-47 fleet, Betsy participated in the North Atlantic crossing, also known as the historic WW2 "Blue Spruce" route to bases in England.
The American contingent, plus other European sister ships, took part in the leading event of the year, which was the "75th Anniversary of D-Day" and took part at Duxford Airfield UK and Caen Carpiquet Airport Normandy France.

"That's All Brother" leading the line-up at Duxford UK, during the 2019 Daks over Normandy event.

The Commemorative Air Force - Central Texas Wing "That's All Brother" N47TB (c/n 12693) rolled of Douglas Aircraft's production line in Oklahoma City, Oklahoma during 1944 and was delivered to the USAAF with s/n 42-92847.

She arrived in the UK, initially for service with the 8th Air Force, on April 26th, 1944, before transfer to the 9th Air Force the following day for service with the 87th Troop Carrier Squadron of the 438th Troop Carrier Group, based at RAF Greenham Common England.

She took part in the late hours of June 5th 1944 and the early morning hours of June 6th dropping paratroopers in Normandy (Operation Overlord). She also took part in Operation Dragoon, July 1944, Operation Market Garden, September 1944, the Siege of Bastogne and Operation Varsity during March 1945. She was declared surplus to requirements and placed up for disposal at Walnut Ridge, Arkansas in October, 1945.

Her first civil registration was NC88874 with Air Cargo Transport Corp in 1946. She went through a number of civilian owners in the following half century before ending up with the Randsburg Corporation of Texas. She was sold Basler Turbo Conversion's, destined for a turbine conversion.

After discovering her true WW2 history, the Commemorative Air Force started a Kickstarter campaign and was successful in acquiring her. Together with Basler they were able to restore her and bring it back into pristine wartime condition.

As part of the D-Day Squadron's Mighty Fifteen DC-3/C-47 fleet, she participated in the North Atlantic crossing also known as the historic WW2 "Blue Spruce" route to bases in England.

The American contingent, plus other European sister ships, took part in the leading event of the year, which was the "75th Anniversary of D-Day", which took part at Duxford Airfield UK and Caen Carpiquet Airport Normandy France.

During my 2015 Californian road-trip I re-visited Chino Airport California and met up with a true WW2 veteran with some incredibly history. This is the **MISSION BOSTON D-DAY** LLC - N62CC Skytrain aka "Virginia Ann" which belongs to Mr Barry Fait who was kind enough to invite me on-board for a local training flight. The a/c name, "Virginia Ann Shoemaker", comes from one of the original crew members wife's name and he took part in D-Day.

She rolled of the Longbeach production line as a regular C-47A-DL with (c/n 13798) for the USAAF with s/n 43-30647. (Sept 1943) She was transferred to the 12th AF/9th AF - 15 Troop Carrier Squadron and took part in Operation Neptune - Mission Boston on D-Day 6th June 1944. She also took part in Operation Varsity - 24 March 1945. As part of a 21 ship fleet with the 59th TCS (X5) she departed Chipping Ongar with British 6th Airborne Division paratroopers to drop

in Germany DZ 'A' near Wesel area of the Rhine.

After WW2 she became NC62570 and in 1955 she was converted to a HIPER DC3-R2000 by Pan American World Airways at Brownville Texas. The conversion included the installation of more powerful Pratt & Whitney R-2000 radial engines, deletion of the LH cargo door and installing large Viewmaster windows. Most HIPER DC-3s went to South America for high altitude operations.

N62CC had a chequered career; back in 1993 it was bought by "Catalina Air Inc" of Alcoa (TN). Catalina Air was a FAR 135 air charter company offering passenger luxury air charters in a corporate configured DC-3. The aircraft offered 16 seats, a full bar, TV, stereo, galley two restrooms, club seating and two side mounted divans. In 2017 she was repainted in olive-drab D-Day colors but its interiors stayed the same.

Top: Flabob California 2017

Left: Duxford UK 2019

Top: French Coast North of Le Havre flying towards Normandy. 2019

Virginia Ann
Normandy
June 6, 1944

As part of the D-Day Squadron's Mighty Fifteen DC-3/C-47 fleet, she participated in the North Atlantic crossing also known as the historic WW2 "Blue Spruce" route to bases in England. The American contingent, plus other European sister ships, took part in the leading event of the year, which was the "75th Anniversary of D-Day". This took part at Duxford Airfield UK and Caen Carpiquet Airport Normandy France.

Bottom, left:
Father and son, Barry and Cole. Duxford

Flabob Express C-47A-30-DL N103NA (c/n 9531) photographed at Flabob Airport CA during an November 2015 visit. Delivered from the Douglas Longbeach production plant for the US Army Airforce with s/n 42-23669 (May 1943), she was transferred to the RAF 24 squadron as FD879.

She was used as a VIP transport for the Royal family and high ranking officers. She operated in India & Pakistan and served with the Air Command S.E. Asia.
After the war she was modified as an executive transport (DC-3C) by specialist converter Remmert-Werner of St. Louis Missouri, with P&W R1830-Super-92 engines (October 1955).

Flabob EXPRESS

Located at historic **Flabob Airport**, Riverside, CA

During her career in Canada she was used as an executive transport with Ontario Paper Company/Dominion Tar/Chemical Co Ltd and Laurentide Aviation ltd with registration C-FIKD. Early 1980s she was converted as a freighter with Ilford Riverton Airways and Air Manitoba.
In 1993 she went to Nostalgic Airways Ltd, Kidron OH with new registration N103NA and Classic Express Airways based at Chino CA.

Currently she is now-known-as "Flabob Express" and based at historic Flabob Airport Riverside CA and is available for Airshows, Scenic rides, Tours and Funeral/memorial fly-overs.

(**Note:** both the a/c data plate and the Remmert-Werner conversion plate show incorrect serial number, not sure how the confusion came about, but the correct serial number is not '33569' but should be '23669'. Possibly a wrong data plate was installed during its time in Pakistan.)

Top: Old Warden UK

The **Aircraft Guaranty Corp Trustee** "101st Airborne Tribute" N150D (c/n 4463) rolled off the Longbeach Douglas production line as a C-47-DL with s/n 41-18401 initially for Pan American July 1942. After WW2, the aircraft was transferred to the French Air Force, joining them on November 20th, 1945 and serving faithfully for the next two decades before being sold on to the Israeli Defence Forces (IDF) in January, 1967 as 4X-FNE/1432 032. There is some indication that the aircraft served with the Ugandan Air Force for a period before her retirement from the IDF in November, 1995.

The aircraft then appears as N155JM on the U.S. civil registry in August 1999, with Global Aircraft Industries of Phoenix, Arizona. Then Ozark Airlines Museum in St Louis, Missouri acquired the aircraft in October, 2001.

Her present owner from Switzerland, Hugo Mathys purchased her in 2016 and she was basically in stock C-47 configuration. The owner commissioned Basler Turbo Conversions to bring the aircraft to Oshkosh for a restoration and have it completed for the D-day 75th anniversary.

The restoration work was extensive, Basler put close to 40,000 man hours into the project. The aircraft went to the paint shop in mid-January, 2019 and it was ready to go by March 7th. Painted in USAAF olive-drab D-Day colours and displaying a new nose art "Airborne Rendezvous with Destiny" under the cockpit window, she was a big hit at Duxford UK and Caen Normandy.

Left: Crew member Carol Voogt and pilot Rebecca Williamson

One of the highlights of the Flabob DC-3 Fly-In was the appearance of the immaculate "**Golden Age Air Tours**" gleaming C-41A N341A (c/n 2145). Normally based at Nut Tree Airport, CA she was a special VIP guest at Flabob Airport, Riverside CA.

This is a very unique aircraft, one of only two C-41s built for the US Army Air Corp. It rolled out of the Santa Monica plant as a DC-3A-253A, but converted to a C-41A for the USAAC 1st Staff Squadron with s/n

40-070. It was converted to a military 14 seat VIP transport and was used by high ranking officers and Major General "Hap" Arnold. After WW2 it was dropped from the USAAF inventory and transferred to the Reconstruction Finance Corp to be sold for civil use (1957).

During its civilian career it operated as an executive transport with a number of oil companies and private companies. She carried numerous registrations such as

N4720V, N65R, N598AR, N32B, N132BB, N132BP and N14RD. During this period she was heavily modified with the AiResearch "Maximizer" speed kit conversion, which included special engine cowlings, engine exhausts, oil cooler fairing, prop spinners and wheel well doors. It increased the cruising speed with 20 miles per hour and its range with 200 miles.

During the early 1980s she was used as flying billboard for a local chicken fast food restaurant. She carried large "Tinsleys Boss Birds" titles on the fuselage and its wings.

Early 2016 she was purchased by its present owner, Aerotechnics Aviation Inc and was refurbished by "Aerometal International" at Aurora Oregon to its current pristine condition. "Golden Age Air Tours" has kept its luxurious 14 seat VIP interior which includes side divans, wood panelling cabinets, storage, VIP toilet, folding tables, and window curtains. In an addition she has a fully functioning air-conditioning system.

Golden Age Air Tours

The **Historic Flight Foundation** (HFF) Longbeach built C-47B-1-DL N877MG (c/n 20806) was delivered to the USAAF back in July 1944 with serial number 43-16340. She served with the China National Aviation Corporation (CNAC) with tail code 100 later as XT-T-20.

With CNAC she served the Hump operation making hundreds of flights over the Himalayan mountains, ferrying supplies and personnel. Flown to Hong Kong at the end of the Chinese Civil War and resumed passenger operations with CAA and Civil Air Transport (CAT).

Back in the US (1953) she was converted to a DC-3C by Grand Central Aircraft Co and was re-registered to N37800. This conversion took 6 months and included upgraded engines, clamp-shell main gear doors, panoramic windows, air stairs, VIP interior and most likely a longer nose cone. Johnson & Johnson took delivery of the

aircraft in October of 1953 and changed the registration to N800J (later as N8009) and flew her as a VIP/Executive transport as seen today.

In 2006 she was bought by Seattle aviation lawyer John T. Sessions who planned to restore her in 1940s Pan American colours. The restoration was completed in 2012 and N877MG was flown to her new home Paine Field with HFF.

Center: Duxford UK 2019

As part of the D-Day Squadron's Mighty Fifteen DC-3/C-47 fleet, she participated in the North Atlantic crossing also known as the historic WW2 "Blue Spruce" route to bases in England and participated in the numerous events at Duxford UK and Carpiquet-Caen Normandy, celebrating the 75th anniversary of D-Day.

(***Note:*** Both Douglas and Grand Central conversion plates show the a/c serial no 4193, which is incorrect, this is most likely a line number)

Another highlight of the Flabob DC-3 Fly-In was the beautiful "Spirit of Benovia" which is a Santa Monica built C-53D-DO Skytrooper N8336C (c/n 7313). Delivered to the USAAF (1942) with serial number 42-47371 it was transferred to the 10th Airforce based at Karachi India. After WW2 it was sold to Central Air Transport Corp (CATC) in Beijing China (1946) and later re-sold to Civil Air Transport (CAT) in Taipei (1949) as N8336C. She was converted by Executive Aircraft Services to a DC3A-SIG3G in June 1955. Going through several owners she finally ended up with Benovia Winery based out of Santa Rosa CA (May 2009).

This is truly a one-of-a-kind beautiful aircraft with a Garrett Corporation: AiResearch Maximizer Speed Kit conversion. Inside the cabin it features a luxurious VIP interior with leather seats & sofa's, panoramic windows, curtains, floor carpeting, wood panelling cabinets/bulkheads, VIP toilet. Externally she features prop spinners, landing gear doors, short stack engine exhaust, tail-wheel fairing and elegant LII passenger entry door. One of the most attractive DC-3 I have ever seen!

To mark the 75th D-Day Anniversary, the "Spirit of Benovia" was repainted in its 1946 Civil Air Transport livery and as such she visited England, France and Germany last year.

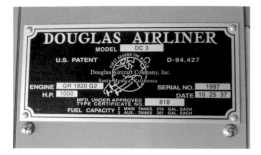

Right: Caen, Normandy 2019

The oldest aircraft flying in the US "D-Day Squadron" was this immaculate Douglas DC-3-201 N18121 (c/n 1997) which is privately owned and registered to **Blue Skies Air LLC**. She set a new Douglas record in being one of the oldest DC-3s to cross the North Atlantic in recent times! She was built for Eastern Airlines as NC18121 ship '341' back in October 1937. During WW2 she was converted to a C-49G and served Patterson AFB with tail code 42-56630.

After the war she was converted to a DC-3A and served with Trans Texas Airways and Provincetown - Boston Airlines/Eastern Express as N136PB flying mainly in South Florida. Between 2010 and 2015 she was based at Aurora Airport Oregon, painted in period Eastern Airline colours.

Current total airframe time is in excess of 91,500 hours... more than 10 years in the air! She is owned by Pete Nickerson and operated by Aerometal International LLC in Aurora Oregon. Her refurbished cabin features leather Gulfstream 3 seats, carpeting, wood panelling, side mounted tables, new galley and lavatory.

One of the D-Day Squadron most attractive members, also known as "Miss Montana", is the "**Museum of Mountain Flying**" N24320 (c/n 20197), she flew across the North Atlantic, during May 2019, in order to take part in the Daks over Normandy

events and celebrated the 75th anniversary of operation 'Overlord', the D-Day landings on June 6th 1944. She was one of the star attractions at Duxford, England and Caen-Carpiquet, Normandy.

She was built as a C-47A-90-DL with USAAF with s/n 43-15731 May 1944; she did not go to war but stayed with in the US. In 1946 she was sold to Johnson Flying Services at Missoula Montana as NC24320. During her live she suffered two major accidents.

On the 22nd of December 1954, due to fuel starvation the crew had to ditch the aircraft on the Monongahela River, 3 km SE of Pittsburgh-Allegheny County Airport, with substantial damage with 10 fatalities out of a total crew of 28. Amazingly the airframe was salvaged and repaired and she operated with Evergreen International, Basler and McNeely Air Charter.

In Dec 15th 1994 she was again involved in a 2nd incident, when a Cessna 208 collided with her on the taxiway at Memphis TN. She was officially retired from active service late 2001 and sold to the Museum of Mountain Flying at Missoula International Airport, Missoula, Montana for static display inside the museum.

Another D-Day Squadron member, who made the North Atlantic crossing during May 2019, was the elegant **Legend Airways** N25641 (c/n 9059) also known as "Lady liberty". She participated in the "Daks over Normandy" events and celebrate the 75th anniversary of operation "Overlord" the D-Day landings on June 6th 1944 Normandy.

She rolled off the Douglas Longbeach production lines as a C-47-DL with serial number 42-32833 February 1943. That same year she was stationed at Oran AFB Algeria North Africa and later in 1944 to the 8th Airforce. Based out of Barkston Heath Airfield, England, she took part in the D-Day troop dropping missions on June 6th 1944 over Normandy France.

After the war she was decommissioned, sold into private hands and was converted to a DC3C-S4C4G by The Dee Howard Company San Antonio Texas (1970). As corporate executive aircraft she led a much pampered life, flying out of Shreveport, Louisiana for nearly 25 years. After numerous operators she ended up with Legend Airways of Colorado LLC, Morrison, Co.

Not flown much she underwent a multi-million dollar overhaul with Basler Turbo Conversions in 1995. Her new corporate VIP cabin interior is nothing short of spectacular and includes a unique 14 club seat configuration, plush floor carpet, window curtains, wooden cabinets, aft lavatory, galley and state-of-the-art avionics.

JB Air Services LLC of Brighton Colorado is the current custodian of N25641. She was one of my personal favorites at Duxford and Caen!

PMDG Flight Operations beautifull N33611 (c/n 17111/34378) aka "Clipper Tabitha May" was delivered to the U.S. Army Air Forces at Douglas Aircraft's plant in Oklahoma City, OK with s/n 45-1108 late 1945. Her military service was brief, with the aircraft going into storage in Augusta, Georgia the following year.

Her initial operator was the Columbia Broadcast System, who registered the aircraft as NC54542 in 1947. She passed through several operators before being donated to the Experimental Aircraft Association in August, 1983, where she was flown by Paul Poberezny and others before her sale to Grand National Air in 1984 as N54542.

She passed through several other civilian carriers, such as California Air Tours and Air Grand canyon Yosemite, before ERA Alaska Airlines bought her in 1995. ERA working with Basler Turbo Conversions in Oshkosh assisted them to rebuild this ship and placed her on the companies Air Carrier Certificate, as a DC-3C.

She was based at Anchorage Alaska, publicized as the last DC-3 to fly in scheduled passenger service in the US. She was known as N1944H "Spirit of Alaska" and performed local pleasure flights from Anchorage, over the scenic landscape and similarly her passengers enjoyed the romance of flying in fabulous 40s and 50s.

She was withdrawn from scheduled service in 2003 and remained for sale until 2011 when she was acquired by PMDG Flight Operations LLC based at Stafford Regional Airport, Fredericksburg-Stafford VA. In 2013 she was painted in full Panam colours with new registration NC33611, as a tribute to the great Pan American World Airways.

Piloted by the Former airline pilot Robert S Randazzo and his gallant crew, Clipper Tabitha May was the first DC-3 to cross the North Atlantic and arrived at Duxford Air Museum UK for the 75th Anniversary of D-Day June 6th 2019.

Note: N33611 was sold to a new operator, called Vearus Jet Sales and will carry the new registration N345VJ

CAF "Inland Empire Wing" C-53D-DO N45366 (c/n 11757) aka "D-Day Doll" was photographed at Flabob during the DC-3 Fly-In. She is painted in a WW2 olive-drab D-Day livery and was initially delivered as a genie C-53D "Skytrooper" for the U.S. Army Air Corps with serial number 42-68830 (July 1943). The C-53s didn't have the large cargo doors and reinforced floors installed. Only 380 C-53s were built during the war. D-Day Doll also sports Wright Cyclone R-1820 engines rather than the more usual Twin Wasps.

She was assigned to the 434th Troop Carrier Group, 72nd Troop Carrier Squadron at RAF Aldermaston, England, and is a veteran of Operations Overlord (D-Day, Normandy France), Market Garden (Holland), Repulse (Bastogne, Belgium) and Varsity (the crossing of the Rhine, Germany).

After a distinguished war career, she was sold to Pen Central Airlines as NC45366 and was converted to a DC3-G202A for civil use. Before American Airpower Heritage Flying Museum, Midland, TX bought her in December 2001, she went through several owners such as Jim Hankin Services and Florida based Shawnee Airlines. Currently she is based at Riverside Airport California and is operated by Commemorative Air Force-Inland Empire Wing.

Together with 14 other D-Day Squadron sister-ships she made the long journey from California to Europe in order to participate in the 75th Anniversary of D-Day, which was celebrated at Duxford UK and Normandy France.

Top: Flabob, Riverside CA
Right page, bottom: Duxford UK 2019

The **Tunison Foundation** "Placid Lassie", formerly known as "Union Jack Dak" C-47A-40-DL N74589 (c/n 9926) is a genuine WW2 combat veteran.

She was built at the Longbeach factory and delivered to the U.S. Army Air Forces with s/n 42-24064 back in July 1943. She was allocated to the 74th Squadron, 434th Troop Carrier Group of IX Troop Carrier Command in England in preparation for D-Day. On the 6th of June 1944 she took part in operation Boston/Chicago over Normandy, towing Waco CG-A4 and Hadrian cargo gliders. There after she participated in a number of combat missions which included: Operation MARKET GARDEN in the Netherlands (September 17- 25), Operation REPULSE – the relief of Bastogne during the Battle of the Bulge (December 23-25) and Operation VARSITY (March 1945).

After the war she was converted to a DC-3C freighter and operated with a number of operators such as: West-Coast Airlines, Aero-Dyne, Saber Aviation, Express Air Cargo and Dodson International Air.
In 2010 she was rescued by James Lyle and Clive Edwards and after an intense eight-week restoration they managed to fly her off to the 75th DC-3 anniversary at Oshkosh. She was painted in blue & white colors with a British flag painted under the cockpit.

Repainted in authentic D-Day colors she crossed the North Atlantic to take part in the 2014 70th Anniversary of D-Day at Cherbourg France. Currently she is registured to the nonprofit organization "Tunison Foundation" and is based at Waterbury Oxford Airport, CT. In May 2017 she flew across the US to take part in the Flabob DC-3 Fly-In.

In May 2019, piloted by owner Eric Zipkin and crew, she flew across the Atlantic Ocean for a 2nd time, on this occasion she lead the D-Day Squadron to Duxford, UK and Caen, Normandy for the 75th Anniversary of D-Day.

Bottom: Weeze, Germany 2019

WRECKS AND RELICS

LA VANGUARDIA - VILLAVICENCIO

COLOMBIA (2015)

Right: Former **Alcom** DC-3 **HK- 4045** (c/n 14363/25808) was built as a C-47B-1-DK with s/n 43-48547 and delivered to the RAF Montreal August 1944 as KJ869. During her military career she was converted to a C-47D and lasted until 1970. She went to the Air Power Museum 1970 as N10004. Sometime later to Air Siesta Inc, Mc Allen Aviation, West Indian Fruit Company, HP-1176CTH Panama Cargo three. She was stored at Opa Locka Florida for a while in 1990. Finally she left Florida for Colombia and became HK-4045x with LAMA at Medellin 1991. Alcom bought this Dak in 1997 for operations from Villavicencio. Presently she is already out of service for more than 17 years and moved from the airport for a new tourist attraction.

Centre, left: Former **Aero Vanguardia** DC 3 **HK-3199** (c/n 14599/26044) was built as a C-47B-5-DK September 1944 for the USAF with serial number 43-48783. This ship is a Vietnam veteran and flew as an AC-47D with three 7.7 mm Vulcan mini-guns! It arrived in Colombia back in April 1976 and flew with SATENA with tail number '1123'. This aircraft had a landing incident at Acaricuari in 2007 while operating for Aero Vanguardia. It was still stored near the

An-12BK at Villavicencio during our visit in April 2015

Bottom, left: **HK-4292** (c/n 13177) C-47B-DK under restoration at the Aliansa maintenance base. Originally she has been manufactured in 1945 at the Oklahoma City plant as a C-47B-DK. This aircraft had two 1200 hp Pratt & Whitney R-1830-92C's engines with

a MTOW of 13.600 kg. This former Vanguardia DC-3 did operate for the USAAF in Australia for a brief period before being sold to Faucett Peru in 1946. From 1967 she served for the Peru Navy. Finally she was imported into Colombia and did operate for AeroVanguardia before being sold to Aliansa.

Bottom: HK 3350

The **Wings of Liberation Museum** Skytrain L4 was photographed, at the museums grounds as part of a crash display back in June 2009. She was constructed as a C-47B-30-DK by Douglas at Oklahoma City with (c/n 16371/33119) for the United States Army Air Force with serial number 44-76787 delivered April 1945. She joined the French AF (Armee de l Air) as '476787', later flew with Air France, Fretair and Ste Uni-Air as F-BAIF.

After her civil career she was displayed in military D-day colors at the Victory Memorial Museum, Arlon in Belgium. In 2002 the Wings of Liberation Museum bought her and transported her to Best Holland. In August 2010 a Dutch Musical Production needed a C47 for the "Soldier of Orange" musical, a WW2 play and was able to release it from the museum. Transportation was needed to bring it to the Valkenburg Theater. During the night transport she got badly damaged when the convoy struck a

bridge, rendering useless for the musical. (ps: a second C-47 had to be sourced).

The severely damaged C-47 was purchased up by Hans Wiesman aka "The Dakota Hunter" from Holland and he recovered the wrecked Skytrain. In 2013, the cockpit was saved and sold to a Dutch Flight Simulator builder based in Utrecht. They reconstructed the C-47 cockpit into the world's first, fully certified - fixed based DC-3 Flight Sim, built on the DDA Classic Airlines Cockpit layout. It is now fully operational and numerous DC-3 pilots across the globe have trained inside this simulator.

Bottom: Stored for many years in the tropical South Florida heat this **Sky Freight** C-47A-20-DK Skytrain N314W (c/n 13041), was delivered to the USAAF in April 1944 with s/n 42-93160, she had served in North Africa with TSP from 1944 until the end of the war in 1945.

Top Right: I am dedicating this book to my friend Hans Wiesman (74) aka 'The Dakota Hunter' He unfortunately passed away last year.

After WW2 she was sold to Harbican & Air Fresh Sea Foods Inc based in Memphis TN as NC8820. She had operated in Canada for many years as a freighter with registration CF-HGD.

Early 1980s she was sold Lee County Mosquito Control District, Fort Myers, Florida. Apparently obsolete and unwanted she was stored and left alone at a deserted corner of the airfield. I photographed her in 1990 at Lee High Acres and she was already in a bad shape, with the RH engine missing, flat tires and damage to her aft flight controls.

Ps: In 2018 she was scrapped and her cockpit was saved and transported to Holland for a private owner.

The ex **Millardair** N587LM C-117D Super Gooney was missing some essential parts when I photographed her on the TMF Inc ramp at Opa Locka Airport (OPF), Florida back in November 2006.
She served as part donor for the TMF fleet, her flying days definitely over her situation got worse, when I re-visited OPF again in 2016, when I discovered that the cockpit was removed and hauled away.

N587LM (c/n 43369) was delivered as a R4D-8 to the USN with BuNo 17108 November 1952. Her military career lasted until May 1978, after which she was stored at Davis Monthan AFB Arizona. Her civil career started a few years later with TBM Inc as N8538F. She migrated to Canada in November 1983 flying for Millardair Ltd as C-GDIK.

Middle & bottom: During a 2013 midwest road trip through the US, We visited "**Dodson Aviation**" Inc at Ottawa Kansas. At that time they were second US Company involved in converting the standard DC-3s into a turbine powered DC-3TP.

During our visit the president, Robert L Dodson gave us a tour of his facility. Two

turbine DC-3s (ZS-OJJ and N81907) and piston-powered N211GB were noted stored in front of the small office. Inside the company's hanger, another turbine DC-3 ZS-OJM was found and as it turned out, she was in the final stages of being delivered to a new customer. Her ultimate destination was a secret, but later we learned that it went to Aliansa SA in Colombia.

After our visit we were granted another tour of Dodson International Parts and Storage yard....just a couple of miles down the road. This is the place where Dodson International keeps a massive inventory of aircraft parts collected over the past 30 years.

The storage area is huge place, but we were mostly interested in the wrecked DC-3. Dodson keeps several containers of DC-3 parts in a back lot. Since a few years it also holds the fuselage and wings of turbine DC-3 N834TP, which was involved in incident in the Mojave Desert. This DC-3 was converted into a turbine conversion and highly modified with a belly radar pod and chin FLIR turret.

N834TP (c/n 12590) serial number 42-92754 is ex KG478 Dakota III RAF Nassau March1944, ex SAAF 6834 with 44 Squadron, Converted to a DC-3TP (C-47TP) December 1997. Based at National Test Pilot School, Mojave Airport, Mojave, CA

as N834TP. She crashed on February 2009 while taking off from Mojave-Kern County Airport, CA. The damage was substantial and spelled the end for this aircraft.

Sadly another ex **TAVIC** C-47B-30-DK, CP-1419 (c/n 16240/32988) in a much worse condition is slowly decaying in the dense air at Cochabamba airport Bolivia. Delivered to the USAAF with s/n 44-76656 she spend most of her time in the military until her sale to Bolivia in 1963 with the Bolivian AF. June 2010

TAVIC - Transportes Aereos Virgen de Carmen C-47A-10-DK CP-607 (c/n 12570) was photographed at Cochabamba "Jorge Wilstermann" Airport Bolivia. She is in fact a WW2 veteran delivered to the USAAF in February 1944 with serial number 42-92736 and transferred to the 8th Europe and 9th AF 34 Troop Carrier Squadron TCS. After the war she became NC19346 with Pan American Grace Airways and later with Lloyd Aero Boliviano (LAB) as CB-34.

Majestic Air Cargo 1944 C-47A-90-DL N67588 (c/n 20536) freighter was involved in a landing accident in May 1998 when due to fuel exhaustion it made a forced landing in marshy terrain near Anchorage International Airport Alaska.

The captain of this flight was also the owner of Majestic Air Cargo, he died later in January 23, 2001 when flying another DC-3, when he crashed into a mountain near Unalaska, AK.

During a 2007 visit I photographed the unfortunate DC-3C freighter along a disused taxi-way at Anchorage Airport. She was badly damaged, missing vital parts, both wings and engines; she still lingered on hoping for better days.

Rolla National airport Vichy, Mo is home to **Baron Aviation Service**'s fleet of wrecked DC-3s. All three were badly damaged by a local tornado back in January 2008.

N47FJ (c/n 9053) was built as a C-47-DL with serial number 42-32827, also known as "Ada Red" she took part in operation Overlord and flew on D-Day, 6th June 1944 carrying 16 paratroopers to Drop Zones in Normandy France. Later on she was involved in Operation Market Garden, 17 September 1944, dropping paratroopers over the South of Holland.

After WW2 she saw service with Texaco Corp and Champion Spark Plugs Corp. She then moved North to Canada and flew with Toronto based Millardair Ltd, as CF-WCM and was converted to an executive passenger a/c including the installation off panoramic windows, a LH passenger door and undercarriage doors. She also flew with Winnipeg based Midwest Aviation Ltd. Back in the US she flew with Baron Air Services Inc based at Vichy MO.

With her glory days far behind, she still lingers on at at Rolla National Airport hoping some day to be rescued and restored as a WW2 veteran. (Postscript: end of 2019 she was saved by Round Engine Aero and trucked away to its new temporarily home).

Below: This 1942 early model DC-3-455 N486C (c/n 6325) was converted to a C-49K with serial number 43-2000 November 1942. After WW2 it operated with Eastern Airlines as ship '380' and with Frontier Airlines as NC19193. She later flew with numerous operators such as Southern Airways, Zantop and Capital as N49F.

According to one off her data plates inside the cockpit she was converted from a C-49K to a DC-3C for executive use, September 1956 by Remmert – Werner Inc at lambert Field St Louis MO.

Her flying career with Baron Aviation Services Inc started in 1974 and she was based at Rolla National Vichy MO.

Rolla National airport Vichy, Mo is home to Baron Aviation Service's fleet of wrecked DC-3s. One of them is an ex FAA Goony bird registred as N51938 (c/n 14511/25956), she rolled of the production line as a R4D-6 for the US Navy with BuNo 17291 to VR-3 Olathe September 1944.

Her military career lasted until 1956 and then she was sold to Alaskan Airlines as N2757A. Several years later she went to the FAA and they converted her into a TC-47J, with tail number N67/N25.

This modification was a standard FAA Type 2 conversion package, for 'flight' inspection duties. The forward bulkhead between the cockpit and cabin was removed. In the forward cabin an airborne calibration and radio navigation station was installed. All electronic radio racks and black boxes were place mid cabin. The aft cabin consisted of crew seats, galley and toilet.

The FAA retired her in February 1973 and sold her to Linn Tech Collage, Linn Mo.

As result of a local tornado, which ripped away her landing gear she laid flat on her belly. Inside the cabin she still featured a complete FAA interior, which now was covered with years off bird droppings. In November 2019, she was saved by Round Engine Aero, which converted the fuselage as rolling truck.

Privately owned C-53D-DO Skytrooper N7500A (c/n 11693) built at the Santa Monica production plant and delivered to the USAAF with serial number 42-68766 April 1943. She served at Oran AFB with the 35th Troop Carrier Squadron and 9th AF/48 Troop Carrier Squadron. She is proper WW2 veteran and took part in operation Market Garden, September 1944 with the 36th Troop Carrier Squadron.

After the war she, was converted to a DC-3A and started flying for Eastern Airlines as NC45332 ship '391'. Between 1952 and 2001 she was went through several operators.

In 2006 she was registered to Jeff Swain and based at Opa Locka Airport FL. She endured some substantial nose and tail damage from Hurricane Wilma. The C-53 was purchased by Frank Moss and he moved to Punta Gorda "Shell Creek" Airfield, FL where she remains in open storage, a sad end to a WW2 veteran.

Left and bottom: Back in 2006 I was working as a Design Engineer at the Airbus factory in Toulouse, France. During my lunchbreak I visited the "Les Ailes Anciennes" Aviation museum, located within the Airbus grounds.
The ex **Fairey Air Surveys Ltd** DC-3 G-ALWC (c/n 13590) was looking very bad when I photographed her.

She rolled of the Douglas production lines as a C-47A-25-DK with serial number 42-93654 during June1944 and transferred to RAF Nassau as KG723 that same month. She joined the AHQ India Communication Squadron that same period.

Her civil career started with Airtech Ltd in January 1950 as G-ALWC. During her time in Britain she flew with Fairey Air Surveys Ltd and Clyde Surveys Ltd. She was sold to Societe Nouvelle Generale Air Services as F-GBOL based at Toulouse-Blagnac. In 1984 she went back to the UK, flying for Visionair International Aviation, again as G-ALWC.

Early 1995 the museum purchased her for public display. Unfortunately she remains in a semi derelict condition, with several parts missing. Not sure if the museum intends to restore her and display it in a more honorable way.

This abandoned DC-3, registered as PT-KUD (c/n 19778) was sitting across the road from the 'Princess Beatrix' Airport, Aruba main terminal. No titles or former registrations were visible. She sat abandoned & unloved in a derelict state for many years ago. According to the records she was left behind most probably after illegal drugs flight in 1982.

This 1944 Skytrain was built as a C-47A-80-DL and delivered to the USAAF with serial 43-15312 to the 8th Air Force, and was transferred to the 9th AF on 26Apr44. At the end of World War II, the Pan American Import-Export Co. acquired the aircraft and registered it as NC54099. During the 1950s she flew with VASP as PP-SQO. After becoming PT-KUD in 1976, she aircraft was registered on March 1983 to Air America Inc. in 2012 she was used in a local monster truck challenge and was completely destroyed.

Date Aruba 17 November 2008

I photographed this brown DC-3 registered N54608, in D-Day markings, coded Z, 315156, on the south side of Miami International airport, next to runway 09R, at the 94th Aero Squadron restaurant, back in November 2006. She was badly damaged by Hurricane Wilma and is probably irreparable and most likely scrapped.
N54608 started live as a C-47A-1-DK with c/n 11903 and was delivered with USAAF serial number 42-92136 back in 1943. That same year she was delivered to the RAF with serial number FL544. After WW2 she flew with BOAC, BEA and Cambrian Airways as a Dakota III as G-AGIP.

In the mid-60s she flew with the Moroccan Air Force as CN-ALI/92136 and returned to the US during the 1970s. In 1984 she went to the 'Yesterdays Air Force Museum' for static display at St Petersburg/Clearwater, Florida

Museum of Alaska Transportation & Industry (MATI), Wasilla, Alaska

This museum is located along the Parks Highway (3) which runs from the state capital Anchorage up North to Fairbanks. It's about an hour's drive from Anchorage.

The Museum of Alaska Transportation and Industry started in 1967 as the 'Air Progress' Museum. It was a small collection of Alaskan transportation artifacts assembled for the centennial celebration of the United States purchasing Alaska from Russia in 1867. The display was on International Airport Road in Anchorage. Fire forced the museum to close its doors in 1973.
In 1976 the museum re-started again under a new name near Palmer Alaska State Fair grounds. MATI moved to its present location in 1992. The museum mission is the collection, conservation, restoration, exhibition, and interpretation of artifacts relating to Alaska's transportation and industrial history.

The collection also includes some rare Alaska airplanes and a couple of vintage airlines such as Fairchild C-123J Provider, Grumman

HU-16C Albatross and two Douglas C-47s. Both C-47s are displayed on the museum outdoor grounds and are looking very tired.

US Air Force Douglas C-47A-80-DL Skytrain "0-315200" (c/n 19666). This aircraft was first delivered to USAAF in February 1944. She took part in Mission "Albany" Norman-

dy 5/6 June 1944 and Operation "Varsity" March 1945. Later it served with 144th Air Transport Squadron (Light) of the Alaska Air National Guard. She served the military all her life and came to the museum in 1961.

The museum 2nd Skytrain was built as a C-47-DL (c/n 4574) and delivered with

serial number 41-18482 in July 1942. She remained in the US and went directly to the CAA as NC5 later changed to N99. She was converted to a DC-3C by the F.A.A. at Oklahoma City, OK in April 1956, and was transferred to the US Forest Service in March 1977 as N101Z. She arrived at the museum in February 1982.

Shell Creek Airpark (F13), is located just a few miles West of Punta Gorda Airport, Florida and is the home of a local Sky Dive club. It is a small uncontrolled grass runway privately owned Mr. Frank Moss. Frank Moss is a familiar name in the propliner world; he has owned several cargo companies using mainly Douglas DC-3, DC-4 and DC-7 aircraft. Currently Shell Creek is the home of several stored DC-3 parts and fuselages.

Center: N133D (c/n 1499) was photographed at Shell Creek FL during my November 2016 visit. N133D was the sixth DC-3 built and is the oldest one on the U.S. Civil Aircraft Register. It was built at the Douglas Santa Monica plant as a DST-G-144 NC16005, completed on 12th July 1936, and delivered to American Airlines 'ship 120' two days later as NC16005 Flagship "Tennessee". When it flew with Wien Alaska Airlines she was converted to a DC-3A. She was operated by Ozark in the 1950 and later with Academy Airlines out of Griffin GA.

During the 1990s she was seen in poor condition missing engines and props, parked in a remote corner of Griffin Spalding County Airport, Griffin GA. Realizing that this was an original DST (Douglas Sleeper Transport) from 1936 the Moss family came to the rescue. After working for many months they managed to fly her out on a ferry permit to Shell Creek where she will undergo a complete restoration.

Top, right: Sadly the 1939 **Aero Libertad** SA DC-3-277A XA-RPN (c/n 2167) was chopped in half and transported to shell

Creek back in 2015. She once flew with American Airlines as NC21768 and many other operators, such as Province-Boston Airlines PBA/Eastern Express N40PB based at Miami International Airport. The owner Frank Moss picked her up from the Basler inventory and will use her as a parts donor for the restoration of his N133D, the oldest existing DC-3 in the world.

The 1942 ex-**Monroe County Mosquito Control** R4D-1 N220GB (c/n 4438) still lingers on at Shell Creek and is For-Sale according her owner Frank Moss.

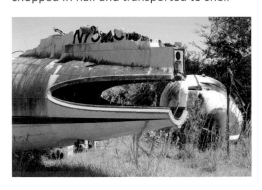

MUSEUMS

March Field Air Museum's Riverside, CA Douglas VC-47A (c/n 20045) with serial number 43-15579 was manufactured by Douglas Aircraft in Long Beach, California and delivered to the Army Air Force on April 12, 1944.

It served in the Air Transport Command before it was transferred to the California Air National Guard. During her military career she was also known by her crew members as "Grizzly".

She was delivered as a C-47A-85-DL but converted to a VC-47A in September 1967. Her military career ended in October 1972. After the acquisition by the museum she was put on display in a basic white and grey livery and dual US Air Force and California Air Guard titles.

The **Airline History Museum** (AHM) 1941 DC-3-362 NC1945 (c/n 3294) was photographed at Kansas City MO during a road trip August 2013. She rolled out of the Santa Monica plant in February 1941 as NC1950.

It was delivered to Transcontinental and Western Airlines (TWA) at Kansas City, Missouri on March 4, 1941 as NC1945 ship

'386'. It served as a civilian airliner through the years with North Central Airlines and with travel club Coronado Airlines from California. She did some geo survey work in Torrance, California, and then sat in the desert west of Palmdale for a few years in storage, as part of an estate. Then in the mid-1980s she went in storage at Roswell, New Mexico.

The former Save-A-Connie group, now known as AHM purchased "old 386" in 1993. It was trucked back to down-town Kansas Airport for a complete restoration.

Seldom photographed is DC-3 EP-TWB (c/n 12680) which is on display at the Tehran **Aerospace Exhibition Center** located next to Mehrabad International Airport Iran.

I photographed her on a cold winter's day. Originally built as a C-47A-15-DK Skytrain for the USAAF with s/n 42-92835 (March 1944) she served with 8th and 9th AF Europe - 61 Troop Carrier Group.

After the war she was bought by Sears Roebuck & Co of Chicago, IL registered as NC49952 and she was known as the "Pacific Coaster" (1951). A couple of years later her registration changed into N78SR. It arrived in Iran with Power & Water Department with new registration EP-TWB March 1972.

This aircraft was converted to a DC-3C by the Garrett Corporation - AiResearch Aviation Service Division with a "Maximizer kit", which included the installation of landing gear doors, short engine exhaust stacks, special engine cowling, oil cooler fairings and a prop spinner. According to the AiResearch brochure this kit added a 20 mph speed increase and 200 miles range. February 2008

(ps: Another DC-3 speed kit which was available, was the Miner's Aircraft & Engine Service Inc. which included special spinners and anti-drag engine cowlings.
It provided more safety, better climb performance, cruising speed & passenger comfort)

Belgium Air Force Douglas C-47B-1-DL, K-16/OT-CWG (c/n 20823), is on display at the **Musee Royale De l'Armee** at Brussels. Delivered with USAAF serial number 43-16357, August 1944, she was transferred to Karachi a month later with the 2nd CCS. She joined the Belgium AF in November 1947.

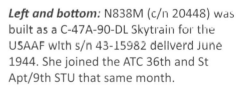

Left and bottom: N838M (c/n 20448) was built as a C-47A-90-DL Skytrain for the USAAF with s/n 43-15982 delivred June 1944. She joined the ATC 36th and St Apt/9th STU that same month.

After the war she was used In the Berlin Airlift operating with the 520th Air Base Group. She was decommissioned in 1974 and transferred to MASDC, Davis-Monthan AFB Arizona for storage. A year later she was bought by Lee High County Mosquito Control Florida and used as a sprayer.

She has become a famous landmark and is displayed as a signpost in front of an "air-roasted" coffee shop. You can't miss "The Roasterie" located at 1204W 27th street Kansas City MO. August 2013

The **China Aviation Museum** is about 40 miles north of Beijing and is located next to the fascinating Xiao Tang Hill (Xiao-TangShan) in Chang Ping County. It is also referred to as the China Air Force Museum by the locals. Part of the museum is located inside the adjacent Datangshan Mountain. There are more than 200 aircraft on display, with an emphasis on the Korean War and the Cold War. This huge facility is the largest of its kind in Asia and is highly recommended for any aviation minded person.

Up to 5 different Lisunov Li-2s and TS-62s can be found on the central diplay area of this grand museum. Former PLA Air Force president Mao Zedong private transport plane Lisunov Li-2 "8205" is displayed near

the entrance of the museum and looks to be in excellent condition.

National Aviation Corporation "XT-115" is a TS-62, a hybrid soviet C-47B conversion, the modification included - ASh-62IR radial engines and extra LH side cockpit window. Some of the museum Li-2s on display have hybrid engines and carry 3 or 4 bladed propellers. October 2012

South Coast Airways C-47A-35-DL G-DAKK (c/n 9798) was photographed on a cold winter day at the Aviodrome Museum, Lelystad Holland. Purchased by **Classic Wings of Holland**, she was dismantled and moved to Weeze Airport Germany late 2013.

This aircraft was deliverd to the USAAF with s/n 42-23936 in July 1943. She moved to Oran 64 Troop Carrier Group in North Africa and was later assigned to the 8th & 9th Airforce European theatre.

She took part in Operation Market Garden (17 September 1944) dropping supplies and troops over Holland. After WW2 she was converted to a DC3C-S1C3G and flew with CSA as OK-WDU/WHA/WZB, French Navy '36' and with Stellair as F-GEOM. In 1994 she was registered as G-DAKK to South Coast Airlines based at Bournemouth International Airport England. Configured with 32 luxury seats he was used for airshow's and provided scenic & corporate flights in classic style. South Coast Airways ceased all operations in July 2002. She was flown

to Holland for temporary storage/display at the museum and possible onwards sale.

Classic Wings wanted to restore her back to former WW-2 glory due to the fact that G-DAKK had operated over Holland during September 1944. After 2-3 years the project stalled and she was sold to the "Overloon War" Museum, based in Holland and was transported to her new location early for restoration and static display. December 2017

Photographed at the Dutch **Aviodrome Museum** in striking orange colours is ex Skyways Ltd Dakota G-AMCA (c/n 16218/32966). She was delivered as C-47B-30-DK with s/n 44-76634 back in March 1945. She got transferred to RAF Montreal unit as KN487 that same year.

Air Atlantique bought her in 1977 and fitted her with an oil pollution spraying kit. After being with-drawn from active service she was sold to the Aviodrome museum Lelystad, Holland and repainted in KLM pre-war orange with fake PH-ALR registration and renamed "Reiger". June 2006

ps: she currently used in a Dutch WW2 musical drama, called "Soldaat van Oranje" as a full size military C-47 Skytrain.

Photographed at the scenic **Flabob Airport** near Riverside California is ex Four Star Aviation Inc freighter which spends most of her career at San Juan Puerto Rico, performing numerous cargo flights to nearby Caribbean Islands.

N131FS (c/n 16172/32920) was originally built as a TC-47B-30-DK for the USAAF with s/n 44-76588 at the end of WW2. She stayed with in the Air Force until September 1970 when she was bought by Desert

Air Parts Inc at Tucson AZ with registration N88916. Three years later she changed hands and became N67PA with Philadelphia Avn Inc. In 1983 she flew to St Thomas VI and joined the Four Star Aviation fleet.

In 2016 she was ferried to Flabob, and mainly used for spares for a sistership. Currently she is on display in front of the Flabob Airport restaurant painted in US AIR FORCE colors and markings.

During a short 2 days layover in Dubai, I was able to photograph the "**El Mahatta" Aviation Museum** Sharjah flagship Gulf Aviation DC-3 G-AMZZ (c/n 12254) inside the spotless building.

According to the 2006 Air Britain DC-3 "Bible", she was delivered with USAAF s/n 42-92452 and transferred to RAF Montreal as FZ669 Feb 1944. She stayed in Canada and went to the Canadian Air Force as CC-129 Dakota with tail number 12943. She operated in Canada with Aero Trades Western Ltd based in Winnipeg as C-GCXE. In 1984 she was sold Haiti as HH-CMG and later as HI-502.
She was withdrawn from use and last seen stored at Opa Locka airport FL as N688EA, Octobers 1999.

Ps: editor's note, there is another DC-3 with fake registration G-AMZZ which is on display at the Natural History Museum (Educational Science Museum) in Kuwait City, which is the former G-AGKE.

The Hellenic Air Force museum
Dekelia Air Base, Tatoi Athens Greece
Date 16 October 2021

The Hellenic Civil Aviation Authority (CAA) Douglas DC-3 SX-ECF (c/n 16458/33206) is preserved at the Hellenic Air Force Museum at Dekelia Air Base, Tatoi Athens Greece. For many years she was displayed in front of the CAA HQ near the old Athens Hellinikon Airport. It was kept for almost 30 years at the Civil Aviation Authority headquarters, after its last flight in 1978 and then moved to the HAF Museum in August 2018. She was built as a TC-47B-35-DK by Douglas at Oklahoma City factory and delivered as USAAF serial 44-76874 and later converted to a R4D-7 with the USN with BuNo 99836. Delivered to the Hellenic Aviation Authority in March 1965 with registration SX-ECF she was used Radio Aids Calibration aircraft for the Hellenic Aviation Authority. Her career spanned from 1965 to 1979.

The other C-47s at the museums are:
49111 C-47B c/n 14927/26372 (in storage)
KK169 C-47B c/n 15415/26860 (in storage)
KJ960 C-47B c/n 14807/26252 (on display)

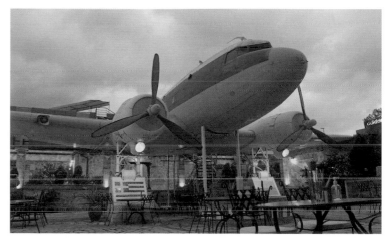

Dakota Beer Pub
33rd street, 13A Elliniko
Near the Athens old Hellinikon Airport

The ex-Hellenic CAA Douglas DC-3 SX-ECD (c/n 14797/26242) is on display inside the Dakota Beer Pub courtyard. Its parts of the restaurant terrace and can be opened up for visitors . Several years ago it was known as the BARIN Russian bar and restaurant, but the new owner has turned it into a vibrant Beer pub and brewery, with the DC-3 as a centerpiece.

She was constructed as a C-47B-10-DK at the Douglas Oklahoma factory and delivered with USAAF serial number 43-48981. Ex KJ950 RAF - Royal Air Force October 1944, later to Hellenic AF. She joined the Hellenic CAA/Ypiresia Politikis Aeroporias in 1958 and was withdrawn from use May 1993 at Athens. After many years of storage at Athens Hellinikon Airport, during October 2005 she was transported by road to her current location.

The **Finnish Aviation Museum** located at Helsinki Airport has two Douglas DC-3s on display.

Top: Douglas Ċ-47A (c/n 19309) with USAAF serial number 42-100846 is located in front of the museum entrance. She was delivered on the 21st of December 1943 and was enlisted into the 8th Air Force March 1944 and later the 9th AF/93 Troop Carrier Squadron. She participated in operation "Market Garden" (Sept 1944). After WW2 she flew with Aero O/Y as OH-LCD, known as "Lokki" and later with the Finnish Air Force as DO-8. After being withdrawn from use in December 1984 she was ferried to Helsinki for static display. OH-LCD is owned by the Airveteran O/Y DC-3 club and is on loan to the museum. She was moved to her current location back in June 2015.

The Finnish Aviation Museum second Douglas Dakota is an early model DC-3A-214 delivered in August 1937 and was transport

overseas and overland to Fokker Aircraft Netherlands for European delivery. Registered as SE-BAC (c/n 1975) she operated with ABA aka "Falken" (Dec1937) and later with SAS aka "Folke Viking". In 1954 she was re-registered as OH-VKB with Karhumaki – Kar-Air and was fitted with geophysical equipment. Withdrawn from use in April 1987 and towed to the Finnish Aviation Museum Helsinki Airport (Sept 1994).

The **Dutch National Military Museum**, located at the former Soesterberg AFB, Holland re-opened early 2015. Its C-47A-90-DL Skytrain (c/n 20118) was repainted in silver Royal Netherlands Air Force (RNLAF) colors with T443 markings and now hangs from the museum ceiling.

It was delivered to the United States Army AF with s/n 43-15652 back in April 1944 and served with the 9th AF, 92 Troop Carrier Squadron TCS in operation Market Garden September 1944.
Its previous identities include 315652 with the Royal Norwegian AF and 68-688/K-688 with the Royal Danish AF.

Center: H-008 was built as a C-47B-10-DK with (c/n 15011/26456) and delivered in November 1944 to the USAAF with s/n 43-49195. She was transferred to the Turkish Air Force as 43-6008, November 1946 and noted with drawn from use during September 1985.

She was put up as display during 1989 as H-008 at the **Istanbul Aviation Museum** (Hava Kuvvetleri Müzesi), located on the South-side of Istanbul Atatürk International airport.

Bottom: YSL-52 was built as a C47A-DL with (c/n 13877) and delivered to the 8th Airforce October 1943 with USAAF s/n 43-30726. During May 1948 it was sold to the Turkish Air Force as 43-6052. Seen wfu during April 1979, then preserved at outstanding Istanbul Aviation Museum (Turkish Airforce) (Hava Kuvvetleri Müzesi) located on the South-side of Istanbul Atatürk International airport since 1985.

Left page and right: During a city trip to Istanbul, I photographed the ex Nesu Air C-47A-20-DK TC-ALI (c/n 12830) aka "Pam-filya" high on its poles, at the Rahmi M KOC Transport Museum along the Golden Hoorn Inlet.
She was delivered to the 8th AF Europe & 9th AF 47 Troop Carrier Squadron with s/n 42-92970 in April 1944.

After the war it was sold to Ford Motor Company/General Motors Corp during June 1946 as NC57779 and converted to a DC-3C-S1C3G. She left the US and was taken up by the Yugoslav Air Force with tail number 71288 (1976). Most of her life, she flew as a corporate aircraft and currently still retains a 70s period VIP interior. March 2011

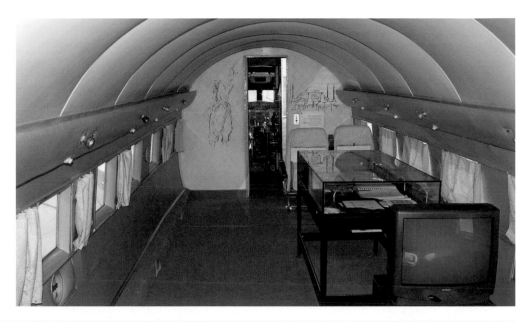

Whenever you visit LAX International Airport Los Angeles, make a stop by the **Flight Path Museum & Learning Center** located next to the Southern runway. It's a great little museum packed with airline artefacts, pictures and memorabilia on LAX Airport history.

Its prized display is the vintage Douglas DC-3-G202A (c/n 3269) ex TWA as NC1944 ship '385' which was delivered January 1941. She retired from TWA during 1950 with a total flight time of 34.259 hours. She was then sold to Pacific Aircraft Sales Company as N1944 and later resold Union Oil Company as N760 and used as a corporate VIP aircraft. Currently she is on loan from the LA Science Museum and as "Spirit of Seventy Six". She on display next to the museum main building. Her VIP interior can be viewed during walking tours organized by the museum. November 2015

Flight Path Museum LAX

Seen here on display at the **Lyon Air Museum**, John Wayne Airport, Orange County Santa Ana CA is   DC-3 NC16005 representing a American Airlines ship "Flagship Orange County". Strangely she never flew with American Airlines. She rolled of the Douglas production line in 1944 as a C-47A-75-DL Skytrain (c/n 19394) delivered to the USAAF with s/n 42-100931. She operated with the 8th & 9th Airforce – 97 Troop Carrier Squadron (TCS) in Europe.

In 1956 she enterd the US register as N56U and was rebuild by Remmert Werner Corporation to a civil DC-3C with an luxurious interior, panoramic windows and uprated engines. In addition a LH passenger door was installed. It was sold to Baldex Corp at Saint Louis MO.

A year later it was sold off to Canada as CF-JUV, flying with numerous operators such as Trans Canada Pipe Lines Ltd, Northern Thunderbirds Air, Knight Air Ltd and North Cariboo Air. She went back to the US register with Spokane based Salair Inc as N394CA.

During the mid 1990s she flew with ERA Classic Airlines aka "Spirit of the North" (N1944M) on sight-seeing flights based at Anchorage International Airport, Alaska. In 2003 she was offered for sale and ferried to Reno Nevada. The newly formed Lyon Air Museum bought it in 2008 and placed her on display a year later.

The **American Airlines CR Smith Museum** at Forth Worth, Texas is the home of the 1940 early model Douglas DC-3-277B NC21798 (c/n 2202). She was originally delivered to American Airlines 'Flagship Knoxville' ship '98' on the 11th of March 1940.

She flew with Colonial Airlines and Eastern Air Lines during the late 1940's and mid-1950s.
Late 1957 she was reregistered to S & W Motor Lines as N393SW and early 1991 the retired aircraft was purchased by The Grey Eagles, an American Airlines retiree group which restored her to its current condition. In November 1992 she was transferred to the museum and was on display in front of the museum and later she moved inside.

The **1st Austrian DC-3 Dakota Club** N86U (c/n 13073) aka "Arizona Lady" was photographed during the International DC-3 Fly In at Salzburg Airport Austria, which took place in July 2014. She was delivered to the USAAF with s/n 42-93189, April 1944. After the war she joined TWA fleet as NC88823 ship '327'. In 1956 she was purchased by oxford Papers & Co and reregistered as N86U.

In 1988 she was ferried to Austria for her new owner Freefall Enterprises.

Several Dakota's took part in a DC-3 meeting organized by Salzburg Airport and the 1st Austrian DC-3 Club, which included Malev Lisunov Li-2T HA-LIX, Swissair DC-3 N431HM, DC-3 Vennerne/Danish Dakota Friends C-47A Skytrain OY-BPB.

CANADA

First Nations Transport Inc (FNT) bare metal DC-3 CF-QHY (c/n 14560/26005) was photographed at her home base at Gimli Municipal Manitoba Canada.

This airframe was delivered by Douglas Aircraft Company with c/n 14560 but due to duplication of serials during production, it was re-serialled with 26005 and delivered as 43-48744 on 16 September 1944.

It was transferred as ship '977' to RCAF 3rd TC Dakota 4ST in September 1944, operated through a series of units & locations. Its tail number changed to '12958', with the Canadian Armed Forces, and served with 402 CFTSD at Saskatoon. After the war she converted to a DC-3C-S1C3G and a Left-hand passenger door was installed.

It entered the civilian register, with Bradley Air Services of Carp Ontario (1971) as CF-QHY, a year later changed hands to one M. Carter of Hay River, NWT. In 1975 it went to Northwest Territorial Airways and after that to Sioux Narrows A/W Ltd of Winnipeg, Manitoba (February 1979).

Plummer's Arctic Lodges of Great Bear Lake became the new owner from February 2002 until 2005.
In 2017 she was snapped up by Basler Turbo Conversions and was ferried her to the company's base for storage.

The sorry looking **Enterprise Airlines** DC-3 CF-OOW (c/n 13342) rolled off the Douglas Aircraft production lines with s/n 42-93431 in May 1944. She was transferred to the RCAF as ship '970', initially with the 3rd TC in June 1944. It served with various units and was re-designated to '12954' in June 1970 and served with 402 CFTSD at Saska-toon. It changed into a civilian hand as CF-OOW for Atlas Aviation Ltd, which bought her on 10 September 1971.

A few years later she flew with Air Dale Ltd, of Sault Ste Marie (1976), Great Northern Freight Forwarding of Ocala FL, Enterprise Air Inc and finally with Triumph Airways Ltd based at Oshawa Ontario Canada (2005). I caught up with her at Gimli Municipal Airfield Manitoba Canada, hiding away in a deserted corner and missing a wheel. In 2012 she was snapped up by Basler Turbo Conversions and was converted to a turbine BT-76 for AIRTEC with registration N167BT.

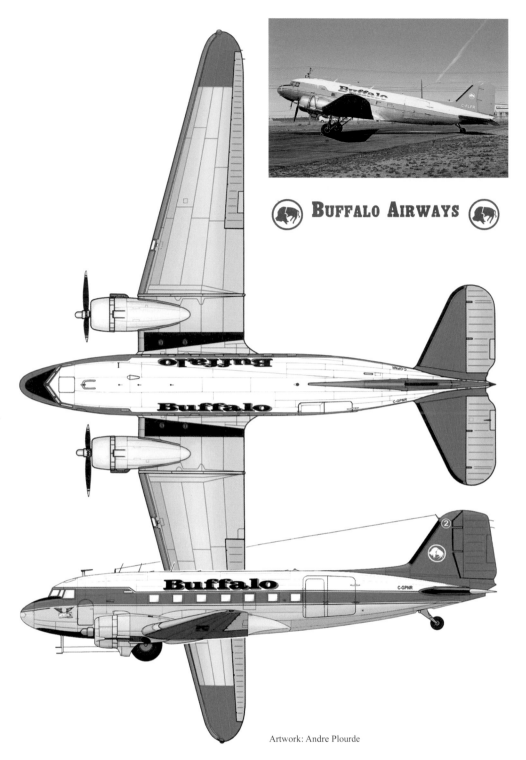

BUFFALO AIRWAYS

Artwork: Andre Plourde

Buffalo Douglas DC-3C C-FLFR (c/n 13155) was photographed on the active ramp, both cargo doors wide open and ready for her next assignment. She was delivered as a C-47A-20-DK early 1944 with USAAF serial number 42-93263. She was transferred to the RAF as a Dakota III with serial number KG563 and joined the 48 Squadron in the UK. After WW2 she arrived back in Canada and flew with St Felicien Air Services LTD and Bradley Air Services.

Bottom, left: Douglas DC-3C C-FCUE (c/n 12983) was photographed in her original 'Department of Transport - DOT' colors and carried 'The Mel Bryan' titles on her nose. She was delivered as a C-47A-20-DK early 1944 with USAAF serial number 42-93108. After the war (1947) she flew with Canadian Pacific Airlines as CF-CUE and later in 1956 she joined the DOT based in Ottawa.

Bottom, right: Buffalo DC-3 C-GPNR (c/n 13333) was photographed on the dusty gravel 'stand-by' ramp. She was delivered as a C-47A-25-DK early 1944 with USAAF serial 42-93423 and immediately joined the RAF Montreal with serial KG602 as a Dakota III. During the 1970s she flew with Canadian AF with serial number '12932' based at Winnipeg.

Buffalo Airways C-47A-5-DL C-GWIR (c/n 9371) was photographed during the early morning hours at Hayriver Airport, NWT Canada, while warming up both Pratt & Whitney R-1830s radial engines.

She is a 1943 Skytrain, delivered with USAAF serial number 42-23509. She was transferred to the 12th Air Force based at Oran, Africa in May 1943. She was involved with the 47th Troop Carrier Squadron, operation Husky and 8th AF/34 Troop Carrier Squadron Sept 1944, Operation Market Garden. After the war she went to the French AF as '23509' and later went civilian with Rousseau Aviation as F-WSGY. In 1975 she arrived in Canada and operated for Aero Trades Western Ltd based in Winnipeg. Several other companies followed such as, Lambair Ltd, Alberta Northern Airlines and Northwest Territorial Airways.

When we arrived in Yellowknife back in the summer of 1998, Buffalo was still operation the world's only regular passenger DC-3 service from Yellowknife to Hayriver across the Great Slave Lake. After meeting up with the owner Joe McBryan for the proper meet and greet, he offered us a return flight on the DC-3. The only thing we had to organize was our overnight stay, which we gladly accepted. The following day we boarded C-GWIR for a 50 minute leisurely flight, BFL128, with Captain Al Fiendel, co-pilot Brian Crocer in the cockpit and F/A Daryl Weber in the cabin, serving nuts and fizzy drinks.

Post script; C-GWIR received heavy damage in a forced landing after take-off at Yellowknife airport back In August 2013 due to engine problems. None of the 24 crew and passengers where hurt. This accident signalled the end of regular passenger flights with the Douglas DC-3.

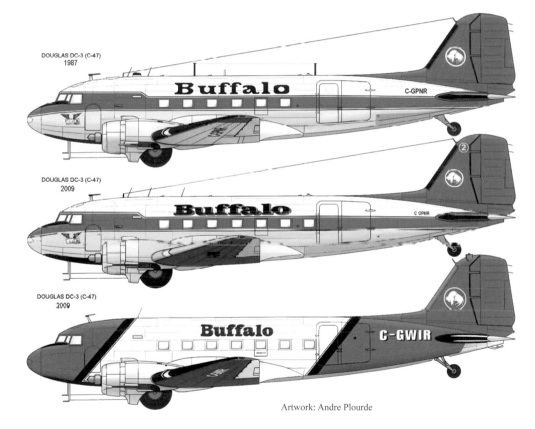

DOUGLAS DC-3 (C-47)
1987

DOUGLAS DC-3 (C-47)
2009

DOUGLAS DC-3 (C-47)
2009

Artwork: Andre Plourde

Amazing Buffalo Airways

"A unique Northern operator in remote regions of Northern Canada"

Buffalo Airways was founded in 1970 by Joe McBryan. He began flying out of Yellowknife, North West Territories (NWT) in Canada during 1961. After gaining his commercial license, he flew for 12 months with Carter Air Service based at Hay River, before enlisting with Gateway Aviation as a bush-pilot flying out of Fort Smith.

Bush-flying in northern Canada in rugged single-engined aircraft is a challenging enterprise, and Joe soon became a seasoned aviator. He joined Great Northern Airways in Whitehorse (Yukon), piloting Douglas DC-3's & DC-4's, thus allowing him to get the feel of these true 'propliner' thoroughbreds. Shortly afterwards he started his own company, which he named 'Buffalo Airways' and

began operations with a single Norseman and Cessna 185. As the business expanded, other types such as Beaver, twin Beech, Navajo and Otter joined the fleet.

The global aviation recession in the late 1970s hit many companies hard, and Buffalo Airways suffered greatly. His company folded and Joe McBryan ended up with a single DC-3. Never a character known to accept defeat, in 1981 he resumed operations with 'Buffalo Air Express' and began an air courier service using a single Travel Air.

He based his rejuvenated company at Hay River and began undertaking priority flights. Slowly the air courier business picked up, and this allowed him to reintroduce the Douglas DC-3 workhorse. The DC-3 proved to be an affordable, reliable and faster platform for his operation, and several other examples were soon added to his modest fleet.

Thereafter, expansion took hold. Joe moved back into air cargo business and renamed his company 'Buffalo Airways Ltd'. In 1993 his fleet comprised five DC-3s and two DC-4s. A year later a new

yellowknife, n.w.t.

type entered the fleet when a single Curtiss C-46 Commando was purchased from Air Manitoba. However, the venerable DC-3 workhorse remained the backbone of McBryan air cargo operations, and Joe developed a special affection for this historic airliner.

By 1994 Buffalo Airways had gathered numerous spare DC-3 and C-47 airframes, which were placed in store at their Hay River maintenance facility. With other companies retiring their examples, Buffalo had become one of the world's largest DC-3 operators. A year later Buffalo was awarded a 5-year fire suppression contract for the Northwest Territories, and began operating 4 CL-215s and 2 converted DC-4s sprayers.

The Government-owned CL-215s retained their standard red and yellow livery, while the DC-4 kept the Buffalo green & white house colors. Both DC-4s have been fitted with a 2200 US gallon fire retardant tank system, installed by Aero Union at Chico, California, and when not used as tankers, both DC-4s were quickly able to make the transition to freighters. Further expansion of the piston-engined fleet came the following year when three PBY-5As Canso amphibians supplemented the fire bomber fleet.

By early 2002, Buffalo's heavy piston fleet consisted of 7 operational DC-3s (plus 5 stored airframes), 3 PBY-5As Cansos, 4 Canadair CL-215's, 2 operational Curtiss C-46s Commandos (plus one additional aircraft expected to be operational in 2002) and 5 Douglas DC-4 Skymasters (including one in storage). In addition, several single engine types were flown, such as the Beech Travel Air, Beech King Air and a single Noorduyn Norseman. During the late 1990s Buffalo added a couple of Lockheed

L-188 Electra freighters for the long haul fuel routes, and retired most of its DC-4 freighters.

Yellowknife, Canada's biggest… little town

Situated at the end of the Mackenzie Highway on the Northern shore of the Great Slave Lake, the ninth largest fresh water lake in the world, "Yellowknife" is the capital city and legislative headquarters for an area touching two oceans and extending over 3.37 million square km, representing one third of Canada's vast landmass. This area is also referred to as the Northwest Territories (NWT). The bustling city of Yellowknife is home to approximately 19,000 inhabitants and lies about 900 km (by air) from Edmonton, Alberta.

Situated just north of the 62nd parallel, the weather in Yellowknife can be very extreme. Highest temperatures are recorded in July, with average temperatures in the low 20C range up to 30C maximum. Winters are long & cold, temperatures can dip down to as low as -30C and -40C in December and January.

Gold put Yellowknife on the map when it was discovered in the 1930s. Then the first settlements appeared, as people of three neighboring tribes moved into the area. Most famous of the mines today are the 'Giant' and 'Con' mines located at either ends of the city. They have been producing gold for almost 50 years. Following the discovery of gem quality diamonds near Lac de Gras, 250 km's northeast of Yellowknife in 1991, the mining business has been booming ever since.

Yellowknife is also a famous tourist attraction, and during the summer months many anglers travel long distances for some spectacular fishing in the Yellowknife River and on the Great Slave Lake. Nahanni National Park is one of Canada's most spectacular natural preserves and popular northern destinations. During the cold winter months many visitors, mainly Japanese, come for the night-time spectacle of 'The Northern Lights', which are most active during the fall and winter. Yellowknife is also the airline hub of the western Arctic, with several carriers such as Air Canada, Canadian North and First Air flying directly to and from Yellowknife.

DC-3 flight to Hay River

During our June 1998 visit we had the opportunity to explore this vast region of Canada. Our main objective was to visit Buffalo's maintenance base at Yellowknife Airport and photograph its hard-working piston-engined fleet. Together with my regular aviation travel friend, Andre van Loon, we purchased low fare tickets flying from Amsterdam Airport routing via London, Toronto and onwards to Edmonton Airport. Most annoyingly, due to a 3 hour 'Air Traffic Control' delays at Schiphol, we missed all our flight connections and arrived far too late for my onward flight to Yellowknife! Fortunately Canadian North came to the rescue, and they arranged for an overnight motel and a much-needed travel & amenity kit, since our bags where temporarily lost in the system.

The following morning suitably refreshed we were re-booked on the midday flight, and resumed our northbound journey. Two hours later we landed smoothly at the tiny airport of Yellowknife, and I savored my first glimpse of a Buffalo Airways Douglas DC-3 parked alongside our Canadian North Boeing B737-200 Combi.

At the baggage claim area we were told that our bags had not arrived on board the same aircraft, but that they were expected within a couple of days! With only our precious camera bags at our side, we rushed outside for a taxi to take us to the centre of Yellowknife, where we checked in at the Bayside Bed & Breakfast located on McDonald Drive in the Old Town area. It is suitable based next to the float plane base. The first settlers arrived in this area back in 1934 and set up transportation links with simple biplane float-planes. Wardair and Canadian Pacific Air Lines started their operations from these lake shores. Today modern bush-planes still dominate the scene, and many local carriers such as Bradley Air Services, Air Tindi, Ptarmigan Air and Air Rainbow can be found, serving distant villages and communities. From our balcony on the first floor we were able to watch the activities of the Beavers and Twin Otters as they landed and took off from the icy waters.

After lunch we returned to the airport for my first introduction to the legendary Buffalo founder Joe McBryan. We met Joe in his small office, which was packed with aviation magazines, photos and memorabilia. We explained the reason

The Bayside

Wayne & Mary Bryant's Bed & Breakfast
3505 McDonald Drive, Old Town
Yellowknife, Northwest Territories X1A 2H2
Phone: (403) 920-4686 Fax: (403) 920-7931
e-mail: becl@internorth.com

* *Located in the heart of scenic historic*
Yellowknife
* *Unparalleled vista of life on*
Yellowknife Bay
* *Within 5 minute walk to all the*
amenities of Old Town
* *Immediate access to the northern*
wilderness out our backdoor
* *Dockside with some mooring facilities*

for our visit and he welcomed us to Yellowknife and most courteously granted us full access to the hangar and facilities.

While discussing the company's current base of operations, Joe invited us to join him on board the daily scheduled DC-3 afternoon flight to Hay River, returning the following morning back to Yellowknife.

Pleasantly surprised by this generous invitation, we gladly accepted his offer and thanked him. This was a unique opportunity, because Buffalo Airways at the time was the only company in the world to offer regular passenger service in a vintage Douglas DC-3! Joe then excused himself and took off in his green and white pick up truck for ongoing business

elsewhere. We walked out over to the oil-stained gravel ramp to start our very own photo tour. Surrounded by DC-3 Dakotas, Consolidated PBY Catalina's, a pair of Curtiss C-46s Commandos and two Douglas DC-4 freighters, we felt as if we had arrived in 'Propliner' El Dorado!

At around 15.00 hours we hitched a ride aboard one of the Buffalo company trucks, which took us to the active ramp in front of the control tower. The immaculate green and white Buffalo flagship Douglas DC-3 C-GWIR stood silently in the glaring summer sun, awaiting its anxious passengers. We were promptly introduced to Captain Al Fiendell and co-pilot Brian Crocer, who was busy refueling the aircraft. After an informal walk around admiring the graceful lines of our vintage DC-3, we then boarded the aircraft making our way through the rear passenger door. Sixteen other passengers soon arrived and joined us in the passenger cabin, for a 50-minute flight to Hay River. I seated myself in the forward cabin and keenly observed the start-up and taxy procedures going on in the cockpit.

ted us to some rudimentary in-flight catering consisting of cool fizzy drinks and a pretzel.

C-GWIR c/n plate revealed it was USAAF Skytrain 42-23509 with c/n (9371) with an delivery date of 13th April 1943. This aircraft was built as a C-47A-5-DL model (manufactured at Long Beach, CA) and served with the 12th AF Oran & 8th AF from March 1944 to September

small passenger terminal, where Captain Fiendell shut down both engines. After disembarking we remained on the ramp assisting the crew in locking up the DC-3 for its overnight stop. We soon discovered why the ramp was so empty when a swarm of vicious mosquitos descended upon the ramp and started biting the recently arrived photographers. We soon decided for a fast getaway and grabbed a taxi for our hotel.

It was a beautiful summer's day with temperatures rising up to 25°C. Several minutes later both engines were fired up and we taxied out to the run-up area. Everything was working fine, and soon we lifted off and climbed to 5,500 feet, thoroughly enjoying the sonorous drone of the Pratt & Whitney R1830 radial engines. Below an impressive lake shore view unfolded beneath the DC-3's massive wings, and at this altitude we were truly able to appreciate the magnificent Canadian landscape as it sped by below. Cruising at 169 knots we flew onwards over the immense lake on course to Hay River. Flight attendant Daryl Weber trea-

1946. She subsequently served with the French AF and Air Attache to Algiers. She arrived in Tucson, Arizona during 1974 and was re-registered to N18262. A year later, Aero Traders Western Ltd bought her as C-GWIR. Subsequently she served with Lambair Ltd, Alberta Northern Airlines, North-West Territorial Airways and finally ended up with Buffalo Airways.

All too soon we approached the southern shore of the Great Slave Lake and the DC-3 lowered its flaps and landing gear. There was no other traffic in the area and we were cleared for a direct approach. After a smooth landing we taxied to the

Postscript:
On the 19th August 2013 DC-3 C-GWIR made a forced landing at Yellowknife Airport. It was on its regular flight to Hay River, and during departure the DC-3 developed engine problems when the RH engine caught fire. The DC-3 circled right for runway 10 then contacted a number of trees, missed some wires and landed very hard and short of the runway, before coming to a stop on its belly. No injuries were sustained, but the aircraft received substantial damage. The a/c is currently in store at Yellowknife awaiting its final fate.

During a trip to Bremen Airport (Germany) for a private visit to the **Alfred Wegener Institute** (AWI) I was able to photograph both Basler BT-67s "Polar 5 & 6" together, while they were in for an annual inspection check. AWI is a global Polar and Marine Research organization and they conduct research in the Arctic, Antarctic and in the high and mid latitude oceans. Both Basler BT-67s are highly modified as research planes and feature an array of probes, antennas, and measuring equipment. (July 2016)

C-GAWI "Polar 5" (c/n 19227) is according to the a/c data-plate, which is located in the cockpit, a DC3-TP67 (50th Basler Conversion) and was initially built from a 1943 piston C-47A. She was modified in accordance with STC-SA00-9. Prior to its conversion she used to operate with Charlotte NC based Saber Cargo Airlines Inc as N79017.

DOUGLAS DC-3 Turboprop Basler BT-67
International Polar Year 2007-2008

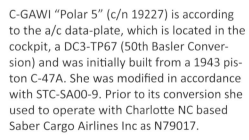

C-GHGF "Polar 6" (c/n 14519/25964) is according to the a/c data-plate a DC3-TP67 (56th Basler Conversion) and was initially built as a 1944 piston C-47B. She was modified in accordance with STC-SA00-9.
Prior to its conversion it flew as a freighter with Miami Valley Aviation Inc as N9923S.

Canadian **First Nations Transportation Inc** (FNT) C-47B-25-DK CF-FTR (c/n 16095/32843) photographed during a cross wind wheel landing at Gimli Manitoba, Canada. Originally delivered to the USAAF with serial number 44-76511 she later joined the RAF Montreal as KN417.

I was able to inspect the aircraft data plates and found that she was converted to a Dakota MK4D by Northwest Industries Limited Edmonton Canada Jan 1953. Later in life she was overhauled and converted to a DC3C-SIC3G by Aero American Corp with US registration N6677. Operating for Rhoades Aviation Inc she was again converted now as a DC3C-R1830-90C.

Two days after our arrival at Gimli June 2009, FNT was shut down by the Canadian CAA and she was put in storage. Buffalo Airways had purchased her and ferried it to Red Deer regional airport Alberta for major inspection. Then in 2016 she was ferried to Basler Turbo Conversions LLC for conversion to a BT-67 standard and was delivered as N144WC.

USA

Atlantic Air Cargo Inc bare metal workhorse C-47A-85-DL N437GB (c/n 19999) was photographed against a typical South Florida background at Opa Locka Airport Florida during a June 2010 stopover. Atlantic Air Cargo is family run business, with two DC-3 freighters under a Part-135 certificate operating daily flights to the Bahamas. President Julio Castrillo also flies his DC-3s together with his two son's Robert and Frank. Delivered to the USAAF in April 1944 with s/n 43-15533 she carried numerous registrations such as NC59360, N222H, VP-BAA, VP-BBL and last HR-LAD with LANSA from Honduras. She was converted to a DC-3C-SIC3G back in February 1946.

Due to engine problems while on approach to Lynden Pindling International Airport, Nassau, Bahamas, Friday 18 October 2019, the aircraft was ditched off the coast. It sank to the bottom of the ocean and luckily Captain Julio Castrillo and his co-pilot were rescued by the Royal Bahamian Defense Force.

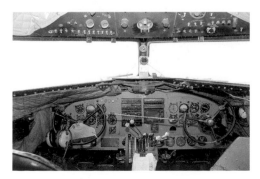

Atlantic Air Cargo Inc other workhorse C-47A-DL N705GB (c/n 13854) rolled off the production line in October 1943 and was transferred to the USAAF with serial number 43-30703. During WW2 she stayed within the USA with several Technical Training Units.

After her civil career spanning from 1945 to 1967 in the US as NX58099, she migrated to Canada with Hudson Bay Oil & Gas Company as CF-HBX based in Calgary Canada. Numerous operators followed such as Arctic Air Ltd, Eastern Provincial Airways, Aero Trades Western and Patricia Air.

During the early 1980s she flew with Transportes Aereos Professionales SA, Guatemala as TG-SAA. In 1985 she came back to the US and flew for J and E Aircraft Inc based at Miami FL. In 1988 she was bought by Atlantic Air Cargo Ltd, a family run business based at Opa Locka Airport Florida. November 2011

Painted in a gleaming polished skin in early Army Air Corps markings, N47HL C-47B-20-DK (c/n 15758/27203) aka "Bluebonnet Belle" belonged to the **American Airpower Heritage Flying Museum Inc** and later to the Commemorative Air Force (CAF), Highland Lakes Squadron.

She was delivered to the USAAF with s/n 43-49942 back in January, 1945 and she served with RAF Montreal and Royal Canadian Air Force as KN270. After WW2 she was enlisted with the Canadian AF as 12909 based at Saskatoon. During 1976 she received a civil registration and served with numerous cargo companies such as Ontario Central Airlines, Nunasi Central Airlines, Perimeter, Calm Air, and Aviation Boreal as C-GCKE. She was photographed during the Warbirds Gathering Show at Topeka KS August 2013.

(Ps: sadly she crashed during take-off from Burnet Municipal Airport Texas on the 21st July 2018. Luckily all crewmembers survived the accident, but the aircraft was lost to impact and fire).

The Ex **Monroe County Mosquito Control** R4D-1 N220GB (c/n 4438) was delivered to USMC MAW-2/VML-252 squadron in June 1942 with BuNo 01982. According to historical records she was involved in the military Guadalcanal campaign between 7 August 1942 and 09 February 1943.

After the war she joined the CAA as N815 and was re-designated as a DC-3A-360. After serving with several companies in the US, she was sold to Monroe County Mosquito Control District, Marathon FL with a new registration N220GB and began a new career as a bug sprayer. (January 1981)

She still lingers on at Shell Creek and is For-Sale according her owner Frank Moss.

TransNorthern
Aircraft Charters

Pictured in this grand over-view at Anchorage International Airport (AK) is **Trans Northern Aviation** "flagship" Douglas Super DC-3 (DC-3S) N30TN. 09 August 2007

Glorious view of TransNorthern "Super" DC-3s line-up at Anchorage International Airport: from left-to-right N851N, N30TN and N28TN, only to been seen in Alaska!

Our Local scenic flight onboard Trans Northers Super DC-3s N30TN was piloted by the Captain Allan Larson (owner) and copilot Robert Hubbard. We lifted off from Anchorage Int Airport and flew towards Palmer, flying allong the Chugash Mountain range, Knik Glacier and over the Matanuska Susitna Valley (Palmer August 2007)

MORE PEOPLE FLY MORE PLACES BY DOUGLAS

Another member of the **TransNorthern fleet** was photographed at Anchorage International Airport AK. The brightly red coloured N28TN (c/n 43354) was converted by Douglas to a USN R4D-8 with BuNo 39080 September 1952. During her military career she was re-designated to a C-117D. Her career ended in 1976 and she went for storage at Davis Monthan AFB Arizona for the next 5 years.

In 1983 she entered US civilian life as N2071X with Reagan Enterprises, Chino and a year later she left for a new career in Canada with Millardair Ltd as C-GGKG and Kenn Borek Air Ltd based in Calgary Alberta. In June 2007 she arrived in Anchorage and joined the TransNorthern fleet.

Photographed at Anchorage International Airport, Alaska is TransNorthern Aviation Inc "flagship" Douglas Super DC-3S N30TN (c/n 43159). She was the 2nd airframe to be converted by Douglas to a DC-3S configuration in October 1949 and carried the registration N30000.

It first flew with Harbet Construction Company Birmingham, AL as N222HC and later sold to Raytheon Company Bedford, MA (1961). This aircraft carried numerous identities over the years including N223R, N6811, N567M and N53315. In 1966 it went to the Mercantile Bank of Monterey in

Mexico where it was registered as XB-NIW, were it stayed for over six years. The aircraft was re-imported privately to the USA as N567M in December 1972 and in August 1976 was sold again to Air Travel Associated in Dallas TX. In the early Eighties it was used as a drug runner on two occasions each time being seized and eventually sold on. It was sold to TransNorthern and moved to Alaska in August 2004.

Left and bottom:
TransNorthern elegant "Super Gooney" N851M (c/n 43302) was converted by Douglas to a R4D-8Z for the US Navy with BuNo 39097. Later in life she was converted to a VC-117D and it flew as a military VIP transport with 16 seats.

After her Navy career which came to an end in 1973, she was declared surplus. She was sold to Florida based Lee High Acres Mosquito Control based at Lehigh Acres, Buckingham, still in full VIP cabin configuration.

In 2009 she moved to Alaska and joined the growing TransNorthern fleet as N29TN for use of flying freight and passenger charters (Part 135). Late 2020 the news arrived that the owner Alan Larson sold to her Alaskan based VIPER as a Veteran transport/maintenance tool.

This was the scene at Hemet Valley airport CA with both **Paralift Inc** bare metal DC-3s getting some maintenance work. This 1939 early model DC-3 G-102A N26MA (c/n 2169) once operated with Penn Central & Capital Airways as NC21781 back in the late 1940s. She was converted to a DC-3-G202A

with Wright 202A engines by Beldex Corp July 1962. In 1977 she was reregistered to N26MA and operated with Parachute Inc and later with Paralift Inc based at Troy MI. Paralift refitted her with Wright R-1820s radial engines.

During her flying career she appeared in two cinema movies: Pearl Harbor & James Bond Quantum of Solace. She has recently been sold to PM Leasing Inc, and is based at Perris Valley airport CA. (Nov 2015)

Sistership N25648 (c/n 2236) was built as genuine DC-3 G-202-A and was delivered to Eastern Airlines, 'ship 355' as NC25648 back in June 1940. She became N20TW with Tri-World Enterprise during 1976 and later on joined the Paralift Inc fleet which refitted her with Wright R-1820s radial engines.

In 2016 she was sold off to a museum in China and was ferried via Anchorage, Alaska and Khabarovsk Russia. She was last seen at Harbin Taiping Airport in the Hellongjiang province, October 2016. According to her crew, she is now parked inside a hangar at Alxa Left Banner Bayanhot Airport, located at in Inner Mongolia, China.

Herpa Wings geniue DC-3A-227 N143D (c/n 2054) was delivered to Swissair as HB-IRO back in October 1938. The Douglas Santa Monica plant delivered this aircraft in major components and shipped it via New York to Fokker Aircraft in Holland, were it was re-assymbled and deliverd it to Swissair. In 1955 she was sold back to the USA as N2817D later to N143D and was converted by Ozark Airlines at Lambert Field St louis, Missouri to a DC-3A with P&W R1830 engines.

During the 1970s she was used as a freighter with Enterprise Flying Services Inc and Jim Hankins Air Services. She lost her passenger interior and the LH passenger door, which was replaced by a C-47 cargo door. During the 1980s she served with Academy Airlines when she got the name "Miss Ali-Gator" due to an accident, after engines problems, she had to make a forced landing in a wildlife refuge at Boca Raton Florida September 1984. She got repaired and remained a freighter based at Griffin Spalding County Airport GA.

She was piloted by Dan Gryder (owner) and James Lewis during an Air-to-Air sortie at the Warbirds & Legends Gathering Forbes Field - Topeka Kansas. August 2013

Ex Skyfreighters and Majestic Leasing Inc genuine "C-47" Skytrain N305SF (c/n 6208) delivered to the USAAF with s/n 41-38749 is a WW2 veteran. She joined the 12th AF ORAN May 1943 and took part in operation Husky with the 29th Troop Carrier Squadron in July 1943. Furthermore she also was involved in Operation Neptune with the 9th Air Force March 1944. After the war she had a chequered career serving in England and Canada. In 1969 she appeared in the US with registration N1157S.

During 1998 she sustained substantial damage during landing at Anchorage International airport Alaska, but she was repaired and ferried to Palmer, where she remained unloved until today.
I photographed her during a cold day with flat tires, battered fuselage and engine cowlings hoping for better days to come. August 2007

One of the Oshkosh 'Air Venture' regular visitors is the **KW Plastic Inc** colorful Douglas DC-3C N728G (c/n 4359). Constructed as a standard C-47-DL Skytrain at the Longbeach production plant (1942), and was delivered with USAAF serial number 41-7860. She did not go to war and stayed in the US and was drafted into the Braniff Airways fleet. After WW2 she started her civilian career with Air Cargo Transport as NC13719.

In 1958 she received a major radio, radar and cabin interior conversion done by Remmert – Werner Inc facility at Lambert Field St Louis MO. In addition special engine cowlings and landing gear doors were added to increase her cruising speed. Numerous operators followed until 1986 when she was purchased by her current owners.

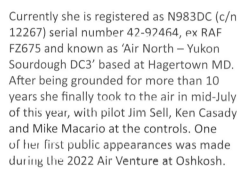

The first time I photographed this 1942 DC-3C was back in 1994 when she was flying with the Canadian operator called **Air North** as CF-OVW aka "Yukon Sourdough". Air North known as the 'Pride of the Yukon', based at Dawson City, Yukon Canada operated 5 DC-3s and a single four engine DC-4 on local routes to Old Crow, White-horse and international routes to Fairbanks and Juneau. All DC-3s were fitted with a lavatory and galley to allow hot meals service for the longer segments. At that time Air North offered a 'Klondike Explorer Pass' allowing travel between five destination for 21 days, priced at CA$550 dollars.

Currently she is registered as N983DC (c/n 12267) serial number 42-92464, ex RAF FZ675 and known as 'Air North – Yukon Sourdough DC3' based at Hagertown MD. After being grounded for more than 10 years she finally took to the air in mid-July of this year, with pilot Jim Sell, Ken Casady and Mike Macario at the controls. One of her first public appearances was made during the 2022 Air Venture at Oshkosh.

Yukon Sourdough is currently owned by a small group of dedicated pilots and mechanics that are very passionate about aviation and preserving aviation heritage for future generations. As a tribute to the original owner, 'Captain Stephen Van Kirk' his name is painted under the left cockpit window.

Florida Air Cargo 2016

My November 2016 South Florida road trip began at Opa Locka Airport for a return visit to Florida Air Cargo Inc. I had visited them many times during the 1980s and 1990s and they have kept operating the iconic Douglas DC-3.

Since February 2012 Florida Air Cargo (FAC) has resumed operations under new ownership with Sergio Raul Alen (President Director of Maintenance), Martin de Urrengoechea (Director of Marketing) and Captain Keith Kearns (Director of Operations). FAC provides Part 135 on demand cargo services to the Bahamas and Caribbean using a fleet of three Douglas DC-3s and a single

Caravan C-208 Cargo master. Due to the built up of the Christmas holidays FAC was extremely busy with daily cargo flights to Nassau Bahamas. During my visit all three DC-3s N15MA (c/n 19286), N271SE (c/n 15676/27121) and N138FS (c/n 9967) were busy fully engaged on the Nassau Bahamas cargo run.

Below: N138FS (c/n 9967) is former San Juan PR based "Four Star Aviation" aka "Snoopy", it had recently undergone a heavy C- Check, which included a wing pull, structural work, new fabrics, replacement of engine control cables, cockpit redone with new instruments and radios. Additionally both P&W R1830s engines and props where overhauled prior to going into service.

Right: N271SE (c/n 15676/27121) ex N300MF is the newest member of the fleet and is the only aircraft that carries the companies name on its fuselage. She rolled of the production line with USAAF serial number 43-49860 January 1945 and was transferred to RAF 300 wing as KN250. Her military career ended in 1950. She stayed in England and flew with Transair Ltd as G-APBC. Numerous operators followed, such as Morton Air Services and Skyways Cargo Airlines. She joined the Missionary fleet in September 1981 as a DC3C-R-1830-90C. She was sold off to Florida Air Cargo in July 2015. Her passenger interior was removed and she currently she is used as a freighter.

Upper Middle/Left:
The tired looking DC-3C-S1C3G freighter N15MA (c/n 19286) was one of the original aircraft that joined the fleet back in 1999. She rolled off the Douglas production line back in December 1943 with USAAF with serial number 42-100823. Early Feb1944 she left for Europe with the 8th/9th Air Force, 91 Troop Carrier Squadron and was involved with Operation Market Garden, September 1944.
After the war she flew with CSA as OK-WDP, French Air Force '100823', Rousseau Aviation with registration F-WSGV. In 1973 she came back to the USA and flew with Mannion Air Charter, based at Ypsilanti, Mi as N15MA. During the 1980s she flew with Century Air Lines and Contract Air Cargo.

Opa Locka based **TMF Aircraft Inc** Super Gooney N32TN (c/n 43301) was photographed, taking off for another cargo flight to the Bahamas. This workhorse used to operate with Alaska based TransNorthern LLC but migrated to South Florida in March 2005.

She was converted by Douglas to a R4D-8Z for the US Navy with BuNo 17175 (October 1951). After her Navy career she was converted to a C-117D and flew with Hawkins & Powers Aviation Inc Graybull, WY as N21270 and later with Lan-Dale as N175TD. TMF Inc acquired her in March 2005 and retained its basic blue and red Douglas Corporation livery.

While in storage at Labelle airport, Florida she was sadly totally destroyed by the Hurrican Irma (2017).

Close up of TMF Aircraft Inc C-117D N32TN fired up both her Wright R-1820-80 nine-cylinder, single-row, air-cooled radial engines. During her time in Florida she still carried the smart Douglas Commercial company livery, which was applied by her previous owner TransNorthern LLC.

TMF Aircraft Inc bare-metal N587M (c/n 43312) "Super" was converted to a R4D-8 for the US Navy with BuNo 12420 back in February 1952. She served with VR-23 Atsugi and later with VR-6 MATS at Westover AFB.

After spending 10 years in dry storage at Davis Monthan AFB Arizona, she entered the civilian market in 1988, and was re-designated as a C-117D, and left for Toronto, Canada flying for Millardair Ltd as C-GJGN.

In 1994 she was purchased by Dragon Air Inc & Air Pony Express. Since my last visit to Opa Locka Airport, TMF Aircraft Inc has gone out of business and its flagship freighter N587M was seen stored awaiting an uncertain future November 2016.

Pro Freight Cargo Services Inc R4D-5 N9382 (c/n 12331) was delivered to the US Navy back in January 1944 with BuNo 17135. After her Navy career, which spanned from 1944 to 1977, she flew offered for sale and flew with Naval Arctic Research Lab from Point Barrow Alaska with civil registration N9382 based at Point Barrow Alaska. (1971)

As a freighter she flew with Northern Air Cargo & IFL Group Inc and finally ending up in South Florida. She was ferried out of Opa Locka Airport to Richards Field for repairs and then transported to the "Days Gone By" Museum in Portland, TN.

Glorious view of the LCMCD piston-engine DC-3 line up Lee County Mosquito Control, Buckingham Field Florida includes: from left to right N838M (c/n 20448) - N836M (c/n 14532/25977) and N211GB (c/n 14688/26133). As off the year 2013 LCMCD retired its piston DC-3 fleet and switched over to the modern and powerful Aircraft Modification Inc (AMI)/Dodson turbine DC-3C-65TP.

Lee County Mosquito Control District (LCMCD) Dodson turbine DC-3 N146RD (c/n 16149/32897) was on display for the general public at Fort Meyers Lee County Page field FL.

She rolled out of the Douglas Oklahoma plant as a C-47B-30-DK with serial number 44-76565 and was assigned to the RNZAF as NZ3538 that same year. Disposed of after the war, she joined New Zealand National Airways in 1947 as ZK AOG/ZK AQP and continued to fly in New Zealand for the next three decades.

She served with the SAAF 44 squadron with tail-code 6858. It was converted to a DC-3C-TP(C-47TP) and operated with the 35 Squadron until late 1990s. Then it was purchased by Dodson International Parts, South Africa (PTY) LTD which was based at Wonderboom Airport Pretoria SA with new registration N146RD. In 2003 she arrived at Dodson International Parts Inc, Rantoul, KS for a conformity inspection and onwards sale.

LCMCD bought her in 2009 and converted her to a sprayer with a set of unique wing mounted spray pods instead of the traditional over the wing spray installation.

Lee County Mosquito Control District (LCMCD) Dodson turbine DC-3 N198RD (c/n 16276/33024) was photographed at her home base at Lehigh Acres, Buckingham Fields Florida.

She rolled out of the Oklahoma plant as a C-47B-30-DK with serial number 44-76692 and was assigned to the RAF Montreal as KN499 April 1945. She stayed in the RAF until December 1955 and then transfer-

red to the German Air Force with tail-code GA+112.

September 1974 she was sold to Fields Aviation (Pty) Ltd in South Africa and she joined the SAAF 44 Squadron with tail-code 6891. The SAAF converted her into a turbine powered a/c using the Aero Modifications 65TP Turbo DC-3 program; generally known as C-47TP.

In early 2000 the SAAF sold her off, with new registration N198RD to Dodson International Parts, South Africa (PTY) LTD which was based at Wonderboom Airport Pretoria SA. In 2007 she arrived at Dodson International Parts Inc, Rantoul, KS for a conformity inspection and onwards sale.

LCMCD bought her in 2011 and converted her to a sprayer with a set of unique wing mounted spray pods instead of the traditional over the wing spray installation.

I photographed this Lee County Mosquito Control District (LCMCD) DC-3 sprayer N834M at her home base at Lehigh Acres, Buckingham Fields Florida. She was on standby status and I was offered the opportunity to view the inside of this 1944 C-47 Skytrain.

N834M with (c/n 14766/26211) was delivered with serial number 43-48950 and was assigned to ATC Morrison October 1944. She also dropped paratroopers during Operation Varsity in March 1945.

During her military career she was converted to a VC-47D with McConnell 23 TFW March 1968.
After storage at Davis Monthan AFB, AZ during 1974 to 1975, she was sold to LCMCD at Fort Meyers Florida and converted to a sprayer. In 2012 she was donated to the Wings of Dreams Aviation Museum FL for static display. Turin Aviation Group struck a deal with the museum and ferried her to Zephyrhills Municipal Airport for restoration to flying status as an authentic WW2 C-47 Skytrain. With new nose-art she became known as "Hit or Miss" including olive drab colors and D-Day invasion stripes.

Lee County Mosquito Control District (LCMCD) 1944 C-47A-90-DL N839M (c/n 20166) is seen taking off for a training flight at Buckingham Field Lee County, Lehigh Acres Florida. She rolled off the Douglas production line with s/n 43-15700 May1944. She did go to Europe at the end of the war based at AFE Rhein Main AFB with 61 Troop Carrier Group. Later during her military career she was converted to a VC-47A and flew with MATS (1970). 1972 she was transferred to U. S. Army Parachute Team,

Fort Bragg NC. Ten years later she was sold off to Lee County Mosquito Control District, Fort Myers, FL.

LCMCD piston-engine DC-3s have since been retired, largely due to increasing cost of maintenance and avgas fuel. The fleet of six were sold off and distributed to museums across the US. N839M went to the Florida Air Museum for static display at Lakeland FL. June 2010

Woods Air Services hard working freighter C-47A-25-DK N50CM (c/n 13445) was photographed against the majestic Chugach Mounting range near Palmer Airport Alaska. Delivered to the USAAF with s/n 42-93524 she joined the 9th Air Force June 1944, she never went to war in Europe. It spend many years in Alaska as a fuel hauler but in 2016 she was snapped up by Preferred Airparts LLC and was ferried to her new home in Kidron Ohio awaiting a possible turbine conversion. August 2007

During its life with Woods Air It was one of the very few US certified "ski-equipped" operational cargo DC-3.

Woods Air Service C-47B-1-DK N777YA (c/n 14189/25634) aka "Arctic Liner" was photographed at the companies home base at Palmer Airport Alaska.
This former US Navy R4D-6 Gooney Bird with BuNo 17259 saw service with VR-3 August 1944 & VR-5 Feb 1945. After the war it flew up to Alaska and operated for Wien Alaska Airlines as N777DG.

While operating a cargo flight for Bush Air Cargo to Nixon Fork Mine AK, November 2015, during landing it struck a snow ridge and the RH landing gear went through the wing before coming to rest. Both crew members remained uninjured but the aircraft was considered a total loss. In October 2021 N777YA was cut up and removed from site.

Missionary Flights International (MFI) turbine DC-3 N200MF (c/n 9766) was originally built as a C-47A-35-DL (1943) with serial number 42-23904 and later transferred to the RAF Middle East as FD933.

In 1945 she joined the South African Air force with tail-code 6879. After her military service she flew with Wonderair as ZS-MRR. During that period she was converted to a C-47TP and as such she flew with Avia Air Charter based at Wonderboom SA. She was transferred to the Kansas based Dodson Aviation Inc fleet as N147RD in 1998.

Bottom and next page top: According to the aircraft data-plate in the cockpit she was converted to a Douglas DC-3C-TP in acc with STC no SA3820SW June 1999. MFI bought her back in 2003 and she was based at the companies HQ at Fort Pierce International Airport, St Lucia FL.

Centre: During my 2010 South Florida road-trip I visited Missionary Flight base at Fort Pierce Airport, at that time they still operated two piston engine DC-3s. N300MF (c/n 15676/27121) sat inside the hangar, I was unaware that she was up for sale.

She rolled of the production line with USAAF serial number 43-49860 January 1945 and was transferred to RAF 300 wing as KN250. Her military career ended in 1950. She stayed in England and flew with Transair Ltd as G-APBC. Numerous operators followed, such as Morton Air Services and Skyways Cargo Airlines. She joined the Missionary fleet in September 1981 as a DC3C-R-1830-90C.

She was sold off to Florida Air Cargo in July 2015 and relocated to Opa Locka Airport, her passenger interior was removed and currently she is used as a freighter.

Missionary Flights International (MFI) piston engine R4D-6 N400MF (c/n 15432/26877) was photographed on the ramp at Fort Pierce airport Florida. Initially built as a C-47B-20-DL at the Douglas Oklahoma production line with s/n 43-49616, she was transferred to the Navy with BuNo 50822 December 1944.

After her Navy career she went on to fly with North Central Airlines as N2400 as a DC3-G202A.
After my visit to the MFI headquarters the company withdrew its piston fleet and N400MF was sold to Kingdom Air Corps Inc based at Palmer airport Alaska.

Missionary Flights International (MFI) at Fort Pierce Airport FL, MFI operated a second turbine DC-3. N500MF with (c/n 15602/27047) it rolled off the production line as C-47B-20-DK with s/n 43-49786. After her military career she was stored at Davis Monthan AFB Arizona awaiting better days.

She was picked up by Miami based Air Sales Inc as N376AS December 1985 and later sold to Aero Modifications International based at Fort Worth Texas July 1990. They converted her into a Schafer DC-3-65TP Cargomaster, which included the installation of two 1,230 hp P&W Canada PT6A-65AR turboprop engines. The a/c fuselage was stretched by 1.02 m forward of the wing root.

She operated some time in South Africa flying with Rossair Contracts P/L as ZS-OBU and a sublease to Sember Aviation AG as 5Y-BNK. She arrived back in the US with Rhoades Partners LLC June 2002.
MFI bought her in early 2010 and ferried her to Fort Pierce Airport FL.

Midlothian, Texas is the home of **Airborne Imaging Inc** and they operate three Douglas DC-3s on a range of special multi sensor missions throughout the US and abroad. All 3 DC-3s have been modified with nose mounted turret and or other radar doors, sensors, camera ports and cabin Equipment racks. The ability to rapidly test multiple configurations on the versatile DC-3 gives

the company 'lab-in-the-sky' concept, according the Airborne Imaging management.

I visited Mid-Way regional Airport, which is located just East of Midlothian, Texas in the fall of 2018 and spotted the highly polished bare metal Airborne Imaging DC-3 N583V (c/n 12369) on the company's small ramp. She was originally built as a C-47A-5-DK

with s/n 42-92555 and rolled of the production line in January 1944, and joined RAF Montreal as KG360 that same year.
After WW2 she arrived in Canada with Canadair Ltd and Imperial Oil Ltd as CF-ESO. In May 1966 she was sold to Artnell Aviation Corp and reregistered to N583V as a DC-3C. She ended up at Forth-Worth Sycamore airfield and spends a total of 30 years in

stored/dormant condition. In 2012 she was made airworthy again and ferried out for a new lease of life with Airborne Imaging. Although built as a military C-47, somewhere along the line she lost her large cargo door which was replaced by a passenger entry door.

Centre: Another member of the Airborne Imaging fleet is the 1942 C-47 N737H (c/n 6062), she rolled of the production line as a R4D-1 for the US NAVY MAW-1 with BuNo 12396. In 1946 she was converted by Remmert - Werner Inc as a DC3C-S1C3G for a private owner with registration NC39340 and was later reregistered to N7H and N737H. She joined the Airborne fleet in August 2002 and was painted in a striking red & white livery which includes a US Air Force roundels on the aft fuselage.

Bottum: 3rd member of the Airborne Imaging fleet is the 1943 C-47-DL N92578 (c/n 9028) which rolled of the Longbeach production line with s/n 42-32802. She did not go to war and spend her military life in the US. After the war she took up civilian markings NC9562H and operated with the Detroit based National Air Transport Corporation. According to one off her data plates in the cockpit she was converted by TIMM Aircraft corp based at Van Nuys California, from a C47 to a DC3C completed in Sept. 1946.

During the 1950s and 1960s she was involved in several US companies with several registrations, such as N75C, N7503, N1800U and N1800D. In March 1973 she left the US and went to Bolivia as CP-1020

and operated by with "Instituto Linguistico de Verano". Eight years later she returned to the USA as N92579 Flying for Missionary Aviation Fellowship out of Redlands California.

During the 1980's she flew with California Air Tours, CA and Nostalgia Air Tours/Island Airlines Hawaii. In 2004 she arrived in Texas and currently flies in a dull two-tone grey color scheme.

The ex-**Frontier Flying Service/Abbe Air Cargo freighter** N59312 (c/n 12363) started its career, like so many other, with a delivery to the United States Army Air Force with s/n 42-92550 on January 1944 and transferred to RAF Montreal as KG354 the next month.

It made the crossing to the England in March where it served with 512 and 437 squadrons. On June 17th 1946 it was registered to the Canadian Government,

which transferred it to the Royal Canadian Air Force (CAF) later that year. It had a long service life with the CAF, but all good things come to an end and it was stored at Saskatoon during 1975.

It entered the Canadian Register as C-GABE, but only to be sold to US Basler Flight Services in May 1978 and reregistered as N59314.

Frontier Flying Service bought it that very same year and brought it to Fairbanks,

Alaska. They were a family owned company founded in 1950 and primarily served the Northern Alaskan community with a Part 121/135 certificate for passengers, freight and mail.

Abbe Air Cargo bought in 2001 and based it at Palmer Airport AK. After many years of open storage at Anchorage International Airport, she was resurrected by to life by Bush Air Cargo and current on lease to Desert Air for the summer of 2020.

This 2-tone camouflaged DC-3, also known as "Lady Luck" was photographed with its wheels sunk in the grass. With her flying days were over, she was slowly gatherings moss on her wings and fuselage.

This 1940 Eastern Airlines DC-3-G202A ship, N408D (c/n 2247) registered as NC15596, was seen sunken knee-deep in the wet grass at the local Sky Dive club at Ottawa Illinois. Her left wing was badly damaged by a tornado back in 2006; she was left unwanted by her owner and became a landmark for the local jumpers and visitors.

The owner, a dentist from Texas, registered to **Big Sky Air Inc**, did inspect his stranded DC-3 and ran the engines occasionally, but apparently over the years he lost interested and it has been sitting on the grass ever since!

Luckily Frank Moss and family came to the rescue and restored her to flying condition. They sourced a airworthy left-hand wing and got the engines running. She was then ferried to Punta Gorda Airport Florida where I managed to photograph her once more. November 2016

PS: When hurricane Ian hit Punta Gorda, Florida this past September 2022, 'Lady Luck' was severely damaged when she was blown away from her parking spot.

The **National Warplane Museum** C-47-DL Skytrain registered as N345AB (c/n 13803) aka "Whiskey Seven" was originally delivered in September 1943 with the USAAF s/n 43-30652. It was transferred to the 12th AF Algiers and later to 79 Troop Carrier Squadron 9th AF 37 Troop Carrier Squadron used in operation Boston. She also took part in operation Market Garden, the liberation of Holland, during September 1944.

It was civilianized in 1945 as NC65135 at Grand Prairie in Texas. In 1963 it was sold off to Canada and as CF-RTB and flew with Eastern Provincial Airways, Gateway Aviation and Skycraft Air Transport. In 2005 she was purchased by the Historical Aircraft Group Museum based at Geneseo NY.

She crossed the North Atlantic Ocean to take part in the 2014 "70th Anniversary" of D-Day, which took place at Cherbourg Normandy France.

The Spotless **Thunderbird Flying Service** C-53D-DO Skytrooper NC43XX (c/n 11665) was photographed at Flabob Airport CA. She rolled of the Santa Monica production line in March 1943 and was taken on Charge with the USAAF serial number 42-68738. She was transferred to 62nd Troop Carrier Squadron, North Africa.

After the war she was converted to a DC-3A by TIMM Aircraft Corp based at Van Nuys California March 1946 and flew with TWA ship '303' as NC86558. Furthermore she went through several owners and registrations, such as N66W and N353MM.

Currently the owner has installed a 16 seat corporate interior with vintage elements such as side-wall divans, Curtains, aft Galley and Lavatory. It's all based on a black & white art-deco theme. She is currently available for charter and pleasure flights under the Wings of Valor LLC group based at Cable airport CA.

Note: NC43XX was sold to a new owner in Brasil and is undergoing full restoration at Aerometal Int.

Since 1990 **Basler Turbo Conversions** has given new life to dozens of DC-3s and C-47s

The Basler BT-67 is a fixed-wing aircraft produced by Basler Turbo Conversions of Oshkosh, Wisconsin. It is built as a retrofit of the original 1940s Douglas DC-3 Dakota and military C-47 airframes, with modifications designed to improve the DC-3's serviceable lifetime for decades to come. The conversion includes fitting the airframe with Pratt & Whitney Canada PT6A-67R turboprop engines, lengthening the fuselage, strengthening the airframe, upgrading the avionics, and making modifications to the wings' leading edge and wing tip.

Basler custom configures each new build to the client's specifications. Industries which have been served include cargo, military, cloud-seeding, scientific research, and various other applications

As close as the original piston DC-3 came to perfection, Mr. Basler knew that there was a need for a better and more efficient version of this highly reliable aircraft. With this vision, Basler Turbo Conversions was created. Production began in the new 75,000 square foot facility in January, 1990 at Wittman Regional Airport in Oshkosh, Wisconsin. Since that time, BT-67 aircraft have been manufactured and sold to customers from every corner of the world. In 1996, control of the company was assumed by Jack Goodale, an aviation minded entrepreneur from Grand Rapids, Michigan. Mr. Goodale brought his considerable skills and background as a builder of businesses to the company and has nurtured and guided the business to its current standing as a complete and focused aircraft manufacturing company.

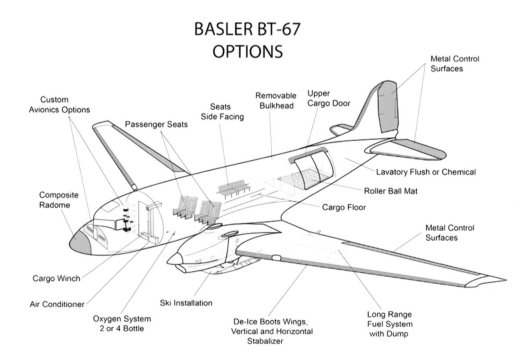

BASLER BT-67 OPTIONS

BT-67 illustration courtesy Basler Conversion

Today, Basler Turbo Conversions and the BT-67 configuration of the DC-3 / C-47 stand as a rare example of successful and complete remanufactured aircraft. The BT-67 and company staff proudly serves a world-wide base of customers.

Info: BT-67 Basler Turbo Conversions LLC - Wikipedia, the free encyclopedia

Note: During an August 2022 visit, yours truly took part on a Basler Facility tour, which was provided by the president Joe Varkoly himself. I was able to photograph all aircraft undergoing conversion and the adjacent storage & bone-yard area.

Basler Turbo Conversions Aircraft Inventory August 2020 - August 2022					
Conv #	c/n	Canadian Reg	US Reg		Location
68	16284/33032		N1350A	Grey colors - near delivery	Hangar
69	12907		N941AT	Ex 'Vera Lynn'	Hanger
70	4654	CF-YQG	N856RB	Bare - ex Nunasi Central	Hanger
71	12438		N700CA	Bare metal with blue cheat-line	Ramp
72	33345/16597		N227GB	Ex CAF 'Black Sparrow'	Hangar
	....	....	....	Bare metal fuselage	Grass
	12597	C-FDTB	N856KB	Ex-Transport Canada	Grass
	43334	n/a	N102BF	R4D-4/C-117D Fuselage only (USMC markings)	Grass
	11625	n/a	N115NA	White with dark blue/red cheatline	Grass
	26005/14560	CF-QHY	N167TW	Bare metal with FNT titles	Grass
	4785	C-FFAY	N856QB	Bare metal	Grass
	9089	CF-JWP	N856YB	Gateway Aviation titles	Grass
	25612/14167	C-GCXD	N856LT	Boreal Aviation titles	Grass
	33170/16422	n/a	N57123	White with light blue cheatline	Grass
	13310	n/a	N115SA	White with dark green cheatline	Grass
	25980/14535	n/a	N68CW	North Cariboo titles	Grass
	27026/15581	C-FQBC	N960BT	Boreal Aviation titles	Grass
	13228	n/a	N843MB	White over faded gray	Grass
	43361	n/a	N100BF	R4D-4/C-117D in the tall weeds	Grass
	9040	n/a	N84KB	White with light blue cheatline	Grass

Note: Parked on the ramp: N300BF Basler demonstrator, N131PR ex AirTec, C-GGSU Xcalibur and 5T-MAH Mauritanian Air Force (for delivery).
Copyright master list: Ralph M Pettersen – updated by M.Prophet (August 2022)

The **Mid America Flight Museum** N5106X (c/n 9058) also known as "Sky King" was built as a C-47-DL. She rolled of the Longbeach production line and was allocated to the U.S. Army Air Force with s/n 42-32832 back in February 1943. She was based at Oran AFB North Africa and joined the 8th AF and later the 53rd Troop Carrier Squadron (3A).

This historic aircraft is a WW-2 combat veteran, she took part in operation Husky/Avalanche (Sicily/Italy), Overlord Overlord/Neptune D-Day (France), Market Garden (Holland) and Varsity (Germany). She mainly flew parachute drop and glider tow missions. After her military career she returned to the Air Deport in San Bernardino, CA for onwards sale.

Top: Close up air-to-air formation with American Flight Museum C-47B N2805J "Spooky" and Mid America Flight Museum C-47 N5106X "Sky King", in a tight formation captured above the skies over Forbes Field Topeka Kansas.

She was converted to a DC-3C with an executive interior which included a LH passenger entry door. She served numerous companies such, General Motors, Inc, Bygone Aviation and Heart of America. She was purchased by Scott Glover in 2000 and he restored it to its current WW2 configuration. Currently based at Mt Pleasant Texas, she visited the Warbirds & Legends Gathering event at Forbes Field Topeka Kansas. August 2013

Desert Air Transport Inc N44587 C-47A-20-DK (c/n 12857) was delivered with s/n 42-92995 to the USAAF on 27 March 1944. She was assigned to the North Africa TSP on April 1944, but returned to the USA the following year. She was transferred to the Reconstruction Finance Corporation (RFC) June 1946 for onwards sale.

She received tail-number NC44587 for West Coast Airlines on November 1946. Many years later it was leased as, CF-ONH for "Pacific Western Airlines" Canada from 1962 and she returned the following year and was re-registered as N44587. Aerodyne Corp, based at Renton, WA became the new owner.

During the late 1970s she was seen in derelict condition at Renton Airport, WA apparently inactive for several years. She was purchased Salair Inc Air Cargo and went back to work as a freighter (1988).
Alta Leasing Inc, Salt Lake City, UT became her new owner in December 1997.
Desert Air bought it from Alta Leasing Inc and ferried her up to Alaska for a new lease of life in the High North.

During an August 2022 visit, Desert Air fleet had grown to four aircraft, since my earlier visit in 2007. The most exciting news was the lease of a Basler BT-67A turbine, N115U (c/n 16819/33567 ex N146Z/N115Z) which arrived in April 2022, which is a first for the Alaskan aviation cargo scene.

Another member of the fleet is DC-3 N272R (c/n 13678), she still retains her longnose radome, from the 1950s when she was operated by Beldex Corp/Westinghouse Electric Corp. The ex-Frontier/Abbe Air Cargo DC-3 N59314 is used as a backup & spares aircraft.

Many thanks to president Joey Benetka, captain Richard Mike Congdom and pilot Cathelin Leoni.

Catalina Flying Boats Inc hard working freighter N403JB (c/n 16943/34202) rolled of the Douglas production line as a C-47B-45-DK Skytrain delivered for the USAAF with s/n 45-940 at the end of WW2. She was converted as a VC-47D and as such, during the 1960s she operated in the US Air Force with tail number 0-50940 and additionally in the Military Air Transport Service-MATS unit as a VIP transport.

She spent several years in storage at Davis Monthan AFB, Arizona before she was sold

off to Desert Air Parts Inc as N17778 (1972). Catalina Flying Boats (CFB) bought her in February 1993 and together with sistership N2298C they operated daily cargo flights to Catalina Island which lies 47 km south-southwest of Long Beach, California.

Catalina Flying Boats (CFB) became the last operational cargo DC-3 operator in the Western US. Both DC-3s were sold to Preferred Air Parts LLC Kidron (OH) in May of 2017 thus ending the Douglas DC-3 legacy in California. November 2015

During my visit at Longbeach Airport, CA DC-3C N2298C (c/n 16453/33201) was already withdrawn from active service. She was stored with her wings & engines removed.

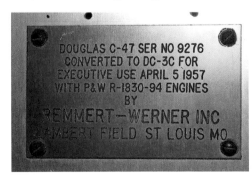

World Jet Inc 1943 C-47A-5-DL Skytrain N8WJ (c/n 9276) was photographed at Fort Lauderdale airport FL, it was offered for sale. During WW2 and after the war she spends some time in Turkey later DHY Turkish Airlines as TC-AFA. She was converted to an executive DC-3C, which included the installation of Pratt & Whitney R1830-94

engines and a VIP interior by Remmert-Werner Inc, St Louis MO April 1957.

It spent most of its life in Mexico and at one time was used by the Mexican Airforce as a VIP executive transport as ETM-6042/FAM-UNO. Early 1993 she was painted in full "Fuerza Aerea Mexicana" scheme and

several years later, Mexicana borrowed her to commemorate the companies 75th anniversary. It was painted in full "Compania Mexicana de Aviacion" livery with a bold XA-CMA registration. It's now part of the Stonehenge Air Museum Inc, Kalispell, MT painted in fake D-Day colors.

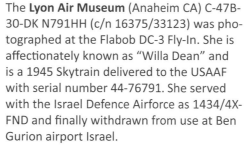

The **Lyon Air Museum** (Anaheim CA) C-47B-30-DK N791HH (c/n 16375/33123) was photographed at the Flabob DC-3 Fly-In. She is affectionately known as "Willa Dean" and is a 1945 Skytrain delivered to the USAAF with serial number 44-76791. She served with the Israel Defence Airforce as 1434/4X-FND and finally withdrawn from use at Ben Gurion airport Israel.

In 2008 she was converted to a DC-3C and was re-registered as N791HH to Cascade Air Inc. She arrived at the museum in August 2006 and painted in a WW2 olive-drab livery including D-Day stripes. She regularly performs at local air-shows in the greater Los Angeles area.

Alaska – The Last 'Propliner' Frontier
TransNorthern LLC

Alaska continues to thrive as the domain of vintage prop-driven transports and it looks like its going to remain that way for some time yet. With this in mind, together with two aviation friends, Chris Mak and Andre van Loon, we made arrangements to visit the 'Last Frontier'. After a grueling trans-Atlantic flight from Amsterdam Schiphol airport, via London Heathrow and San Francisco, we finally arrived at Anchorage 'Ted Stevens' International airport.

Landing just after midnight and extremely exhausted, we picked up our rental car and drove straight to our motel, thankfully only 5 minutes away! We had booked rooms at the Puffin Inn, the name of a well-known Alaskan sea bird. Although our main reason to visit Alaska was to photograph the hard-working aircraft used by the cargo and fuel hauling companies that still use the venerable Douglas DC-4, DC-6, and Curtiss C-46 'propliners', we also decided to visit the famous Lake Hood seaplane base and checked out the local float plane scene.

Glacier Ride
On a previous trip to Alaska in 2003, I met the friendly owners of TransNorthern LLC (TN), Alan and Andrea Larson. At that time they were extremely busy setting up their first Douglas C-117D "Super Gooney" registered N32TN into regular cargo service, and I was fortunate to catch the aircraft at Palmer on its first revenue flight. Alan Larson, who also flies as Captain and his crew where extremely busy, starting up the company. So it was not possible to catch a ride on the aircraft due to FAA monitoring operations. But Alan did extend the invitation to fly with him next time I would visit Alaska!

Based on the Northern shores of Lake Hood, TransNorthern operates a small fleet of aircraft on passenger and freight air charter services to numerous fishing lodges and small villages within Alaska. Four years later, a few things had changed as TN had purchased a new aircraft in the shape of a 1949 vintage Douglas Super DC-3 N30TN, while sistership N32TN had left the company and sold to a company in South Florida. In addition, they had expanded their fleet with the addition of two further C-117Ds, the ex Kenn Borek C-GGKG, which had been re-registered as N28TN, and the ex-Le High Acres Mosquito Control sprayer N851M. A quick look inside the latter revealed a partial VIP interior which included folding tables and brown leather seats. Why this aircraft ended up on the TransNorthern ramp remained a bit of a mystery, maybe it had been acquired as a spares ship.

Finalizing our private charter was an informal event. We discussed our options with General Manager Andrea Larson, and she agreed on a late afternoon flight to nearby Palmer and return at a basic one hour flat rate of $1600, not counting the time needed for pre-flight activities. The aircraft would be parked at gate A7, just across from the old Reeve Aleutian hangar, which was TN's regular departure gate for its normal charter operations near the main terminal.

We joined the crew at the TN office and our co-pilot/flight engineer Robert Hubbart, looking very much like a 1960's hippie, took us out to the ramp. Being a 'round engine' buff himself he gave us the grand tour of the airport, which was much appreciated.

We were fortunate to have excellent weather, and enjoying the moment I pondered on the beautiful looks of Trans Northern new flagship N30TN, which is truly a historic transport. Originally built as a C-50 model back in 1941, she was immediately drafted into the USAAF, but three years later was sold to American Airlines and converted to a passenger aircraft. She never enjoyed a long career with American, as she was quickly sold back to the Douglas Aircraft Corp for the Super DC-3 program, and was converted as the 2nd Super DC-3. N30TN differs from the first Super DC-3 in having a standard left-hand passenger door, Pratt & Whitney R-2000-D7 engines, (instead of the Wright R-1820s), and a 31-seat airline configuration. In total, only 5 Super DC-3 models were built for civilian use and N30TN is the only survivor today!

I followed Robert out on the ramp and joined him on his walk round inspection, and watched as he crawled out on the wings for the oil and fuel checks.

Soon after, Alan joined us and we were all set to go. We boarded the aircraft and it was nice to have the whole aircraft to ourselves. Alan and Robert settled in the cockpit and began going through the 'pre-start up' and 'engine start' check list. Both R-2000 engines fired up smoothly and settled into a rhythmic hum. Once the engine run-ups and checks had been completed, Captain Alan taxied the Super DC-3 off the parking ramp and requested the Echo-intersection for departure from Anchorage's runway 32.

ATC responded quickly, advising that a runway 32 departure would entail a 20-minute wait due to numerous "heavy jets" lining up for take-off.   We had no choice but to abandon this idea, and opted instead for the more standard runway 07 departure. In contrast, we immediately received our clearance and taxied out for a rolling take-off.

ning mountain backdrop. This was going to be our comfort stop and photo-shoot moment with the "Grand Old Lady".

Palmer Municipal Airport is a relatively tranquil place well known for its local residents in the shape of two potentially airworthy C-119 Flying Boxcars. They have been worked on for more than a decade in the hope that they might fly one day! In addition, several Bush Air Cargo DC-3s could be found undergoing maintenance or in storage. On the north side of the airport is an Air Tanker base which is used by Alaskan Department of Forestry, and more recently Alaskan Air Fuel has taken up residence there with two DC-4 fuel tankers.

With just 5 souls on board and no cargo, we quickly lifted off into the breezy Alaskan sky. Gazing out of one of the forward cabin windows I saw the familiar Northern Air DC-6 ramp below, and in the far distance Lake Hood with its abundant floatplanes. A quick scan in the cockpit revealed a steady climb at 95 knots, and the engines running at 2,200 rpm, with a 35 inch of manifold pressure power setting.

Shortly afterwards the power was eased back as we levelled out at an altitude of 3,500 feet, and looking forward I could see the impressive Chugach mountain range approaching and looming large in the cockpit windows. Some minutes later, now settled in a comfortable cruise, a gradual left-hand turn took us over Anchorage's northern city limits. Following the Glenn Highway we had the towering peaks of the Chugach range on our right wing – the highest being Mt. Marcus Baker at 13,176 feet and well above the level of our flight today. Twenty minutes into the flight we approached the Matanuska-Susitna Valley, which lies just south of the city of Palmer. We made a gradual right turn and followed the course of the Knik River, and although the weather deteriorated, with some clouds drifting overhead, we managed to remain in VFR conditions.

Alan and Robert pointed towards the distant Knik Glacier appearing on the horizon. This spectacular glacier averages about 200 feet in depth - its top face is about three miles long at the head of Knik River, about five miles long along the gorge and 3.5 miles long in the valley of Lake George.

We dropped down low for a closer look! The Knik Glacier is part of the massive Colombia Glaciers complex, and we were looking at only a small part of it. The crew really enjoyed themselves as they made a steep low-level pass over Inner Lake George. Below the DC-3's massive wings, several big chunks of iceberg flashed by. We made a 360° turn and back-tracked the blue and white glacier trail as it was time to head out of the valley and land at Palmer Airport, with its charming atmosphere and stun-

After spending an hour on the ground, making the most of the gorgeous summer weather and the spectacular backdrop, we jumped on board our private Super Dakota again for the flight back to Anchorage. The wind was calm, so there was no real need to backtrack for take-off. Both engines were fired up in a characteristic cloud of smoke, and we took off from runway 34, followed by a gradual left-hand turn.

Soon after, we approached the Alaskan Transportation Museum, near Wasilla, with its small aviation collection. We circled the museum twice in honor of the two C-47s Skytrains on display at the museum!

We levelled out on a general heading towards Anchorage International, and after contacting approach control, were vectored for a visual landing on runway 7R, just a few miles behind a Dynasty cargo 747. Wake turbulence was carefully avoided by remaining just a bit higher than the approach path of the heavy jet. We landed long, and with no need for even touching the brakes, we left the runway via the last exit. This time we taxied back to the TransNorthern maintenance ramp and the engines were shut down ending another exciting day in Alaska!

I would like to thank Alan & Andrea Larson and Robert Hubbart for all their help and the amazing flight.

Palmer municipal airport is a small public airport located 2 miles Southeast from Palmer city center, located in the Matanuska Susitna Borough Alaska.

It is situated along the Matanuska River and surrounded by the snowcapped 'Chugash Mountain' range, with Mount Baker towering at 3,991m (13.094 feet).

It's a typical small Alaskan airport with a large fleet of private planes and it also has a Division of Forestry & Fire Protection on the North End with a fleet of fire bombers stationed during the summer months.

It's also the home of **Golden Era Aviation**, which started 2 years ago with a fleet of two vintage passenger Douglas DC-3s

(N400MF and N763A). Golden Era Aviation was created by two local DC-3 pilots and provides a one-of-a-kind vintage flight experience, which includes onboard breakfast and champagne, all served by a stewardess in 1950s period uniform.

Golden Era Aviation slogan reads: Step onboard our vintage Douglas DC-3 and you will be transported back to the 1940s and soar over stunning Alaskan Mountains and Glacier scenery such as the grand Knik Glacier! Onboard there is plenty of time to visit the cockpit and take pictures.

Currently Golden Era Aviation offers two kinds of tours: a 30 minute introductory flight and a 60 minute deluxe flight. Both DC-3s are stationed at Palmer airport, which is only a short drive from Anchorage the state capitol.

Left page: I would like to thank Nico Von Pronay (owner/pilot), Timothy Oldenkamp (owner/co-pilot) and F/A Estelle Strawn for a wonderful flight. August 2022.

Center: Knik Glacier, Alaska

Built in 1944, N400MF was commissioned to the United States Army Air Forces, but was transferred to the Navy as a R4D-6. After over a decade of military service she was purchased by North Central Airlines in 1956, where she flew passengers on commercial flights for many years.

In 1984 N400MF was called to a new line of work, when she was acquired by Missionary Fights and Services. Over the next 30 years, she flew all over the Caribbean for relief missions in disaster and poverty-stricken areas. Even though she had more to give, she eventually was replaced by more modern turbine equipment that phased out these older piston powered aircraft and Kingdom Air Corps, another missionary service, gave her a new home in Palmer,

Alaska. This versatile aircraft can haul passengers or cargo and will hopefully still be able to support relief missions when she is on her breaks from flying flightseeing tours.

One of the goals for this special aircraft, that has yet to be realized, is following the old land lease route that was used to ferry aircraft to Europe during WWII. Now that N400MF has come to reside with Golden Era Aviation, maybe she will get to make this voyage yet!

Golden Era Aviation
Vintage Flight Tours
907-230-8204   231-225-8847
info@goldeneraaviation.com
www.goldeneraaviation.com
820 E Aircraft RD STE 100 - Palmer, AK 99645

N763A, was one of 219 military C-53s (equivalent to a DC-3A) built by Douglas Aircraft at the Santa Monica plant for WWII. Completed on March 11, 1942 she was transferred from United States Army Air Forces possession to the Navy. After being decommissioned from military service she was snapped up by Continental Airlines in the spring of 1947 and converted for use as a commercial airliner. Southern Airways bought her in 1949 for Southeastern U.S. and Central American routes, where she flew passengers until being sold to a Texas business man in 1966.

She has seen the U.S. and Central America as a commercial passenger plane, was bought and sold by a handful of private owners, was overhauled in classic 1950s Ozark Air Lines colors for their 35th anniversary open house in 1985, and made an appearance in many airshows; before settling down as someone's personal plane in Florida. In 2016, after this semi-retirement, we rescued N763A out of Florida (she would have been destroyed by a hurricane a year

later) and flew her all the way to her new home in Alaska.
This aircraft is on the registry of historic places of Illinois and is truly a time capsule. While N763A has carried a lot of people and cargo around the world, probably her

most famous cargo was Grace Kelly, Princess Grace of Monaco, who once flew in this very aircraft.

Info: Golden Era Aviation

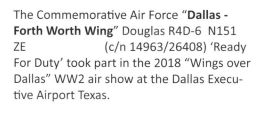

The Commemorative Air Force "**Dallas - Forth Worth Wing**" Douglas R4D-6 N151 ZE (c/n 14963/26408) 'Ready For Duty' took part in the 2018 "Wings over Dallas" WW2 air show at the Dallas Executive Airport Texas.

I have photographed her before when she visited Eindhoven Holland, back in July 1985 for the 50th Anniversary of the Douglas DC-3.

She was built as a C-47B-10-DK with s/n 43-49147 but converted to a R4D-6 with BuNo 50783 and taken up by the USN VR-3 Olathe squadron in October 1944. After leaving the Navy in 1958 she was purchased by the US Department of Agriculture Forestry Service with registration N151Z. She joined the CAF back in 1983 and calls Lancaster Airport Dallas her home.

SOUTH AFRICA

Johannesburg Rand Airport was the home of "**Springbok Classic Air**", fleet which includes the spotless 1943 C-47A-1-DK ZS-NTE (c/n 11926). This aircraft rolled off the Douglas Aircraft Corp plant at Oklahoma under Contract number W535-AC-2405 with a USAAF serial number 42-92157.

It went to the RAF as FL565 and transferred to RAF Middle East on the 20th of January 1944 as a Douglas "Dakota" Mk III. During the war she operated some time in western Libya, Morocco, Italy and Algiers under the SAAF wings as ship '6873'. After the war it was placed in temporary storage which lasted from 1947 until 1960, after which she was made airworthy again for the 86 Multi Engine Conversion Unit at AFB Bloemspruit, Bloemfontein.

In 1995 she was sold off as ZS-NTE and it was ferried to Lubumbashi in the Democratic Republic of Congo, where it was based for some time. It operated cargo flights in the Eastern DRC.

In 2002 it was sold at auction to Captain Flippie Vermeulen, who commenced a full restoration which started in April 2009. The conversion started at Springbok Aviation Services to flying condition in August 2009 in hangar 5 at Rand Airport and was completed in August 2010.

After my visit in March 2015, Flippie Vermeulen sold her to a German businessman, Mr Peter Adrian based in Trier, Germany and ferried it to Europe during May 2015. As off mid 2017 ZS-NTE was repainted in 1960's Aer Lingus colors and carries the historic EI-ACD Irish registration and it is based at Zweibrücken Germany.

Note: In 2020 ZS-NTE was sold to Morlock Aviation as N249CM

SkyClass Aviation Dakota ZS-BXF (c/n 12107) was built as a C-47A-1-DK at the Oklahoma City production plant and delivered with s/n 42-92320 November 1943. It served with the RAF Nassau as FZ572 and later with SAAF 6821/6888 code. In 1948 she then joined the South African Airways as ZS-BXF "Klapper-kop". She rejoined the SAAF until 1972 and served with 44 Squadron. All SAAF Dakotas were soundproofed and fitted out with 21 seats for airline service.

It joined the SAA Historic flight back in 1991 and was based at Rand airport. Klapperkop was painted to represent the DC-3s that began early service with South African Airways during the 1940s. Her main task nowadays is a leisurely one, taking passengers on low-level nostalgia trips in the Johannesburg area.

In March 2015 I re-visited Rand airport located on the outskirts of Johannesburg and spend some time with SkyClass Aviation. At the time they operated the SAA historic fleet of two Douglas DC-4s and DC-3 for passenger/cargo charters, airshow & pleasure flights.

I would like to thank SkyClass Aviation for granting me ramp access and making this photoshoot possible. In addition also a special thanks to SkyClass models "Farryn Ter-Ossepiantz" and "Lezinka de Meyer" for making it an enjoyable event under hot & humid South African sun.

SKYCLASS AVIATION
"Flying on silver wings"

This ex South African Airforce C-47 No 6850 was photographed at the **Johannesburg War Museum**, back in March 2015. She was built in 1945 in the US by the Douglas Aircraft Corporation at Oklahoma City as a C-47B with (c/n 15654/27099) with serial number "43-49838". She was delivered to RAF Nassau as KN231. It was assigned to Air Command South East Asia and went to India.

After the war, in the early 50s she was converted to a Dakota Mk III and sold to Sudan Airways as SN-AAH. After her military career she was sold to the Caesars Palace casino and was put on display inside the building. Currently she is depicted in its Second World War RAF colors and markings.

If you visit Johannesburg Rand Airport, you can hardly miss the **Phoebus Apollo** DC-3 nose section sticking over the fence along the entrance road into the airport. She is on display next to Phoebus Apollo main ramp and hanger.

She was constructed as a C-47B-5-DK by Douglas at Oklahoma City, Oklahoma with (c/n 14483/25928) with USAAF serial number 43-48667. She was delivered to the Royal Air Force with s/n KJ897 as a Dakota III.

After the war she flew with Field Aviation/ BEA as G-MAKE. She served with Rhodesian Air Services as VP-YUU. While serving with Zambia Airways as 9J-RDR she was damaged beyond repair after a ground loop, seen wfu Lusaka in 1994.

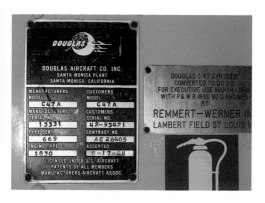

Rovos Air immaculate C-47A-25-DK ZS-CRV (c/n 13331) also known as "Delany" was delivered with s/n 42-93421 May 1944 and transferred to RAF Montreal as KG600, re-designated to a Dakota MK111 that same month. During the war she took part in operation Market Garden (17 September) and operation Varsity, which was the air-borne crossing of the Rhine, towing gliders. Her military career ended in 1953 and she was sold to Bendex Corporation in the USA as N96U.

This is a well-travelled DC-3 with countless flight-hours across the globe in USA, Canada, Cuba, England, Zimbabwe, Botswana, Namibia and South Africa. She passed through numerous owners and registrations such as CU-P702, CU-N702, N702S, G-ASDX, G-AJRY, ZS-PTG, A2-ACG and finally ZS-CVR.

During a scheduled flight between Sishen and Johannesburg for United Airlines, 15 October 1975 a baby girl was born on board. The baby was called "Delaney" hence the name of the aircraft. In June of 2002 she was bought by Rohan Vos of "Rovos Rail" and he overhauled and gave it a complete refurbishment, which included 21 comfortable leather business seats. After a short stay in Namibia with NCA she returned to South Africa wearing Classic Air Travel titles.

During a 2004 visit, the author was lucky to enjoy a breakfast flight on-board "Delany" from Lanseria Airport South Africa. The flight was organized by AIR ROUTE Adventures and gave its passengers a unique flying experience on-board a vintage airliner, with champagne served during the flight. Our 40 minute scenic flight routed over Johannesburg and the surrounded country side.

She was photographed again; a couple of years later in 2015, during a follow-up visit to Wonderboom Airport, SA and she was now registered to the Air-Team group.

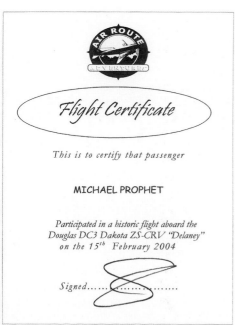

Phoebus Apollo Aviation DC-3 ZS-DIW (c/n 11991) named "Pegasus" was photographed at her home base at Rand Airport Johannesburg during a 2015 visit.

She was delivered with serial number 42-92215 October 1943 and transferred to the RAF Middle East as FL583 later that year Dec1943. In 1944 she joined the SAAF with s/n 6871. After the war she obtained a civil registration ZS-DIW with Aircraft Operating Company.
During her civil career she was operated by South West Airways, Aircraft Operating Company Ltd, Anglo American Ltd and finally Namib Air Registered December 1978.

Phoebus Apollo purchased her late 1990 and used her local cargo flights, with regular service to Gaborone Botswana, mainly with general cargo.

During my 2015 visit at Wonderboom Airport South Africa, I photographed the ex, **Wonder Air** DC-3 ZS-DRJ (c/n 12026) with a new coat of fresh paint. After being flown from Rand Airport to Wonderboom back in 2010, she was slowly being restored. I was able to view the cabin and she featured a complete passenger interior which included an aft toilet.

She was built as an C-47 Skytrain, in October 1943 with a USAAF serial number 42-92247 and went to the Royal Canadian Air Force as FL615. After the war she started a new career with Trans Canada Air Lines as CF-TDO. In 1963 she was sold to Commercial Air Services, Comair as ZS-DRJ.
Not sure what the current plans are..?

SOUTH AMERICA

Lineas Aereas Canedo (LAC) spotless R4D-8 CP-2421 (c/n 43365) was converted by Douglas for the US Navy with BuNo 17190 in November 1952. The Navy converted it to a C-117D configuration back in 1961 while flying for USMC at Beaufort.

In 1983 she flew with Hawkins & Power Aviation Inc as N4504W. She migrated to the Bahamas with Marine Products and later served with Trado as HI-545CT from the Dominican Republic. Afterwards it was resold as N545CT, with MG Marketing Enterprises based in Miami South Florida.

In August 2002 Lineas Aereos Canedo purchased this C-117D and registered it CP-2421. For awhile she was used for nostalgic flights within Bolivia in corporation with Aerosur. She was completely refurbished to carry 31 passengers in a very comfortable cabin with leather seats, hat racks, a lavatory and a small galley. Cool drinks and meals were served by Aerosur flight attendants in classic 50's style uniforms.

The Canedo was a family run business since 1978, based at Cochabamba Jorge Wilstermann International Airport. Due to political and economical unrest in Bolivia the company has since ceased all flying activities and its flagship Super DC-3 (C-117D) was stored at Cochabamba and offered For Sale!

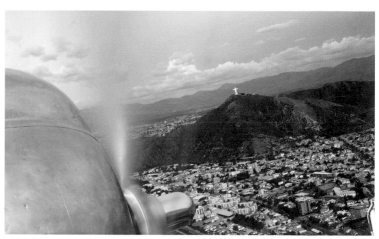

South America revisited

In pursuit of a Bolivian Super Dak

Almost a year had passed since we said goodbye to our friends at Lineas Aereas Canedo (LAC) in Cochabamba Bolivia. We had planned to fly aboard LAC flagship 'Super DC-3' CP-2421 which was operating weekly tourist flights to Uyuni and Rurrenabaque. However, due to an engine failure this flight never took place, and since then a lot had happened in Bolivia. LAC had been supplied with poor quality aviation gasoline, effectively destroying two Wright Cyclone R-1830 engines, at a cost of around $50.000 each! This forced LAC to discontinue the tourist flights on behalf of Aerosur. Meanwhile, Lloyd Aero Boliviana (LAB) one of the oldest airlines in the world, still operating Boeing Boeing 727 tri-jets, had run into financial difficulty. The owner of LAB had been accused of fraud, and was no longer allowed to leave the country.

Within the past year, Bolivia had chosen a new president, Evo Morales - the first indigenous president in the country's history. Some of his more radical ideas had placed Bolivia on the "to be watched" list in the USA. Despite all these tumultuous events we decided to re-visit the highlands of Cochabamba and check on our friends in Cochabamba.

Last year our outbound flight was plagued by cold weather and heavy snowfall, which covered much of Holland, leading to the cancellation of many flights. This year looked like an exact repeat. The original plan was to meet up at Paris CDG airport for the Air France Boeing 747 flight to Caracas. My KLM flight went smoothly and I expected to meet my travel companion 'Andre' at the Air France gate, but he was nowhere to be found.

Shortly afterwards I received a text message with the message, "Unable to reach Paris, flight cancelled due to broken down toilets!" All hell broke loose at the other end, because this would threaten his entire South America trip. I departed to Caracas without knowing if Andre would make it to Bolivia. The following afternoon, I arrived aboard LAB's single airworthy Boeing B727-100 series jet at Cochabamba 'Jorge Wilsterman' airport, via stops at Bogotá and Santa Cruz. Once outside the airplane I sensed the pleasant hot and dry Bolivian air, reminding me of our previous visit. Down on the ramp I recognized a familiar face; it was Roberto Canedo (LAC C-117 copilot and General Manager), who had

come to meet his two photographer friends from Holland. "Hey", he asked, "where is Señor Andre?" I told him that it was a long story, which I would tell him over a nice cool Bolivian Taquina beer.

72 hours later, not having slept much, and having travelled via Milan, Caracas, Sao Paulo and Santa Cruz, Andre arrived at our Gran Ambassador hotel. Right….. finally our holiday could begin. The following day we caught a taxi and directed the driver to the now familiar location of the LAC base. Not much had changed since our last visit; DC-3 CP-1128 was now parked outside the hangar, with restoration work now on hold. Curtiss C-46 Commando CP-973 had moved a mere 30 metres, but looked really astonishing! Entirely polished and repainted, the C-46 looked brand new! She had been completely refurbished to full passenger configuration including comfortable leather DC-6 seats, plush carpet, overhead luggage racks, aft chemical toilet and a small galley. Basically, the C46 was ready for flight. Only some minor work had to undertaken. Even a delay of one year had not brought us any closer to seeing this aircraft in action! According to the LAC family, a test flight was expected soon.

Flagship Super DC-3(C117D) CP-2421 stood in front of the hangar, now devoid of Aerosur titles. LAC had overcome the bad fuel problems and installed a different carburettor, able to handle the higher lead content of Bolivian avgas. Now with the aircraft back in service the

tourist flights had restarted with a once a week charter. The plan was to undertake more flights directly chartered by the travel agents. Unfortunately for us, no flights were scheduled during the week of our stay! Instead we moved to Plan B, and considered the cost of chartering the C-117 ourselves! Before our visit we contacted Roberto about this possibility, and asked for a quote. Of course, this proved to be no problem. Roberto's father and founder of LAC; Rolando Canedo Lopez and Chief Commander of the C-117D, responded by inviting us to detail our requests and asking when we wanted to fly! And so it was agreed that we would fly the very next day.

Bolivian Super DC-3 charter flight

The day of our flight began with a layer of cloud lying over the city and surrounding mountains. It was still early morning, so we decided to shift the flight into the afternoon. The weather had been excellent during the past day with crystal clear blue skies and temperatures in the 30

degrees C. We arrived at the small LAC office just before noon and the whole Canedo family, including older brother Capt: Marcello Canedo (ex LAB F27/B707 and A310 pilot), were attired in full flight gear. We discussed the weather and Roberto assured us that the cloud base would burn off after midday! Activity on the LAC ramp picked up pace. Some of the mechanics were busy polishing the engine cowlings and cleaning the main gear assemblies. I joined Roberto for his external walk around inspection. Then we boarded the aircraft and went inside the cockpit, together with Captain Marcello to go through the 'before start-up checklists. At 12.30 local it was all quiet again and we all went out for the usual feast of a mighty Bolivian lunch! We arrived back at 1400 hours local and activity moved into a higher gear. Fuel was checked and both props where manually turned over. Some last minute preparation, and it was time for a crew picture in front of the imposing nose of the C-117. Rolando indicated that it was time to run the engines. Andre and myself stayed

outside to catch the magical moment of engine start on pictures and video! Both Wright R-1820s radial engines burst into action and started running smoothly as the warm oil lubricated the engines. Having enjoyed the moment for some minutes, we then boarded the Super DC-3.

Seated on the left was Chief Commander Rolando Canedo Lopez, while sat on the right was Commander Marcello Canedo Valdivia and co-pilot Roberto occupied the third cockpit seat. Back in the cabin were seated chief mechanic Victor Hugo Galindo and two cheerful Dutch photographers. We taxied along cross runway-22, runway 32 towards stand 1 on the main ramp. Here the engines were shut down and Roberto headed to the terminal to file our flight plan and pay the landing fee... a total of 109 Bolivianos!

(Roughly $13 dollars) As our proposed flight-plan also called for a low pass over the runway, special approval was requested and granted by the DGAC (Bolivian CAA). Soon Roberto returned with the necessary papers and we were good-to-go! Once again both engines were fired up again and we backtracked across runway 22 towards our engine run-up position. With the brakes firmly set, the Super DC-3 started vibrating heavily, while both engines running at maximum power settings had their magnetos and propeller feathering systems tested. All checks successfully completed, and with a thumbs up we taxied onto the runway. I looked out through the forward cockpit windows, which seemed filled by the soaring mountains at the other end. This was surely the moment when the performance of the Wright engines was critical, and they did not miss a beat. With an empty weight of 21,490 pounds and zero flaps we lifted off the runway and

climbed to 10,500 ft. The engines where performing outstandingly and we made a gentle right-hand turn for a base-leg approach towards runway 22. For the benefit of one more Canedo brother Lalo (currently a Boeing 737 pilot) and LAC mechanics, we made a low pass over Runway 22. We executed a second right-hand turn and this time we levelled out over the city of Cochabamba, thereby marking the official start of our aerial tourist flight. We passed alongside Cochabamba's most famous landmark, the 40 m high Cristo de la Concordia, which is located atop a 265 m high hill overlooking the city. Then we made a left-hand turn towards the Cordillera Occidental range. Here you can find the high peaks of the Tunari National Park. These mountains are very close to the city and both the Aerosur and LAB pilots have to make a steep climb and sharp turn towards the west in order to avoid these towering peaks. But today we are flying alongside these magnificent mountains at a relaxed cruising speed of 160 knots and an altitude of 10,700 ft. We turned to the left and headed south towards the Angostura Lagoon, overflying the little towns of Punata and Tarata, where we saw an emergency strip by the name of Santa Lucia. Unexpectantly Captain Rolando invited me to sit in the right-hand seat and even handed over the controls of the CP-2421 to me. For me this was one of the highlights of our trip! Then it was time to turn back, and we made a 180 degrees turn towards the city. With a firm touchdown on runway 22, this marked the end of our unique and thoroughly enjoyable Bolivian charter flight. So perhaps we should come back and charter the C-46!!!

En-route to Gran Roque

Several days later we boarded another LAB Boeing 727 flight, this time across the great Amazon jungle towards Caracas Maiquetia International Airport. Unfortunately, we failed to see anything of this magnificent jungle since this was a night flight out of Santa Cruz Viru Viru Airport. Our scheduled time of arrival at Caracas was an unsocial 02.10 hours local. Our flight was running late, and this meant an even later arrival. We decided to hang about in the terminal until day break; it's just too dangerous to take a taxi in the middle of the night, especially to our hotel which was in a rough neighborhood, so we had heard. After arrival we looked for a quiet corner and tried to catch up on some sleep!

The main reason we included Venezuela in our itinerary was linked to the fact

that currently no less than two companies operate the faithful DC-3 on daily scheduled passenger flights to the Los Roques Archipelago. Both 'Aeroejecutivos' and 'Sol de America' have an early morning flight (08.00 hours) from Caracas Domestic Terminal directly to Gran Roque, the main island.

Two more companies fly this popular route; Transaven, with its small LET410 and Linea Turistica Aerotuy (LTA) with their De Havilland Dash-7-102s. Located some 168 km (approx: 100 miles) North of La Guaira (mainland) the Los Roques Archipelago consist of some 42 coral islands. The largest is called Gran Roque, which is similar to a big rock. This is where most of the population lives. It also boasts the longest airstrip (800m) and is the destination for most flights. Approximately 1,500 people inhabit the Islands and about 60,000 visitors arrive

yearly. The Venezuelan Government declared Los Roques a national park in 1972. The Island has a wide variety of seabirds and sea life. It is often referred to as the 'Bonefish Paradise'. Many tourists come for scuba diving, go fishing or simply come for the peaceful, beautiful lagoons and sandy white beaches. For us modern day DC-3 prop-hunters, the existence of daily DC-3 passenger operations, coupled with the relaxed atmosphere for photography lured us to this Paradise island.

At daybreak we struggled to get up and walked over to the domestic terminal, which was a 15-20 minute walk, while dragging all our luggage and fending off the usual army of money-seeking "hustlers". At the far end of the terminal we found the tiny LTA and Aeroejecutivos offices on the ground floor. Both were still closed. The only activity was at the Sol de America small check-in counter. This airline derives the name "Sol de America", not as you might expect from the ever-present Caribbean sun, but from Simon Bolivar, the liberator of South America - it was his monicker. Already some tourists were checking-in for the Britten-Norman BN-2A MK3 Trislander and DC-3 flight. After the morning rush, we inquired about the subsequent DC-3 flights. This was not a problem. "When do you want to go?" It turned out to be

pretty much straightforward. Once you have decided when you want to travel, a handwritten ticket was made out. The only requirement was to pay with cash at the counter, with the round-trip fare costing us $ 123 dollars. We furthermore checked in at the Aeroejecutivos (AE) desk, but unfortunately discovered that their DC-3 was engaged in a 4 day charter to Canaima. A new management at AE had come to power and this caused some fleet changes. Their single Convair 440 (YV-223C), which had only flown sporadically, had since our last visit in Sept 1999, been broken up. DC-3 YV-500C was currently operational on passenger flights, while former Colombian DC-3 YV-1179C was acting as a back-up. This latter DC-3 carries a very different colour scheme, having a polished appearance with a yellow cheat-line, although no titles are carried. The former AE flagship DC-3A YV-440C never left Venezuela; the deal fell through and it was left to sit quietly at Caracas downtown airport. Exciting news came from AE chief pilot Omar Pinoso, who mentioned that AE had bought a Colombian DC-6 freighter, which was still at parked

at Villavicencio, Colombia. Already in full colours the aircraft was awaiting a ferry permit from Bogotá.

Two days later we returned to the domestic terminal and checked in for our flight. Check-in procedures were very basic and simple, with not a computer in sight. It was a 500 metre walk to our departure gate, where we first caught a glimpse of Sol de America vintage YV-911C Dakota (32761). The domestic ramp brought back memories of the golden years of first generation 'jetliners'. The ramp was bustling with old jets which where being used for internal and a number of international flights. Companies such as Rutaca and Avior Airlines operated the classic B737-200 series twin jet, while Aserca and Aeropostal were still using Douglas DC-9 series 31 and 51 on their main routes. Santa Barbara Airlines operated a single B727-200 aircraft, and if you still wanted to fly a short body Douglas DC-9-15 series, LASER was the company for you. (A footnote: all these wonderful jets are banned from European airspace for being too noisy!)
In the midst of all this we boarded the Sol de America mini bus, which shuttled us to our brightly coloured, blue-yellow and white DC-3. On the ramp we were greeted by the DC-3 crew, comprising Captain Hernan Marchan, co-pilot Oscar Lizardi and F/A Denial Pestana. Having captured a few ground pictures, we promptly boarded the Dakota as it was time to depart. Our DC-3 turned out to be a vintage model; its c/n plate revealed a manufacture dated from 1944! Researching its construction number I found out that she was delivered to the USAAF in March 1945 with tail number 44-76429.

Officially built as a C-47B-25-DK model at the Oklahoma City production plant. After her military career she was overhauled by Field Aircraft Services and flown by numerous, mainly European operators, including Aigle-Azur, Aer Lingus, Sata Air Acores and Lansa. More recently she was used by a Venezuelan company called Caribbean Flights, based out of Valencia.

Sol de America launched operations in 1980 in Ciudad Bolivar, on the shores of the Orinoco River, and later moved to Margarita Island (1985) by founder member Capt. Hector Freites, an important union leader. Later sold to investors from Caracas in 1999, the airline was brought to Maiquetía, Simón Bolívar International Airport. In 2001 Sol de America was acquired by Chairman Capt. Dror Kassab Ruben, an industrial engineering graduate from Cornell University, and a former crew member on VIASA DC-8/63 and DC-10/30 jets. At the time of our visit the fleet consisted of 2 Let410 UVP, one BN-2A Trislander, and one DC-3. According to Rafael Ruiz, Director of Operations and former KLM and Viasa pilot, two Jetstream J-31s were planned to join the fleet in 2006. He also gave me some interesting background information about Sol America Airways chief pilot Heran Marchen.

Our DC-3 Captain, Hernán José Marchán was an exceptionally experienced pilot. He had flown the following aircraft: Beechcraft Mentor T34; Douglas DC-3/C-47; Curtiss C-46; Convair 340, 440 and Allison-powered 580; Lear Jet; Douglas DC-9; King Air 90-100-200-350; Douglas DC-3 TP (Turbo Prop); West

Wind 1124; Falcon 50 - acting as flight instructor in most of them. He started his airline career in 1961 with Aerovías Venezolanas (AVENSA) and in 1976 he then joined PDVSA (Petróleos de Venezuela) the national oil company, where he served as flight instructor, chief pilot and chief of operations until his retirement in 2002. Rafael Ruiz contacted him in 2004 and asked him if he would be interested in wearing his gloves again. He said "naaaah!, it has been two years and I am taking it easy, I have a good pension and I am enjoying wearing slippers. Nothing bothers me, only maybe that Anita (his wife) has got me watching daily TV soap operas and sending me for groceries too often, other than that, I'm cool". Rafael knew him quite well, having served as his co-pilot between 1972 and 1975. He knew that this guy loved airplanes and could fly them very well too. I told him, "Hernán, we got us a DC-3, 1944 latest issue, with R-1830 engines, we are flying it to Los Roques, how about it?". He came to my office the very next day, and shortly afterwards he was back flying and loving every moment of it, and even using his great experience to train new copilots. He is 64. Total flying time: 30.500 hours and counting.

very spacious with the left side of the bulkhead removed providing a nice open plan arrangement. Both pilots seemed very keen to show us around and to impress us with their professionalism and easy manner with which they handled this big tail-dragger. After the normal engine run-up, we lined up on runway 08. In front of us an Aserca DC-9 twin jet blasted its way down the runway. Now it was our turn. In front of us lay 3,000 metres of concrete and the blue Caribbean beyond. Hernan slowly and evenly opened up the throttles and we were off. After a few seconds the tail went up and shortly after we were flying. Gear-up, power reduced and the DC-3 was trimmed for a leisurely 110kts climb speed to our cruising level of 5,000 ft. We made a left-hand turn and crossed the rugged coastline. A few early morning clouds, about to be burned off by the sun, were still around, but the Dakota sliced through them with ease. The whole experience was utterly smooth and soon afterwards most passengers were asleep, no doubt helped by the steady heavy beat of the engines sending deep vibrations through the airframe. Back in the cockpit all was relaxed as well, as this well-oiled machine made its slow progress towards

relished the panoramic picture post card view below. A string of small islands came into view, indicating we were approaching the archipelago.

A few miles out, the gear and flaps were extended and we lined up for a landing on the short concrete strip that calls itself Los Roques airport. A wary eye must be kept on the numerous yachts and powerboats that cruise the coastline in front of the airstrip. As the runway is only 800 metres long, we came in low and slow, crossed the threshold at no more than 10 feet and taking great care to avoid the new mobile control tower that had been sited on the field. It seemed as if our wing would clip it, but all was well… A healthy crosswind was blowing from the right, so some differential power was used to keep the plane straight. In the flare, as Hernan slowly came back on the power, he kicked in some rudder, dropped the right wing and we made a perfect smooth touchdown that could hardly be felt. In fact, some of the sleeping passengers in the back were hardly disturbed from their slumbers. He made it seem all too easy. More than 35 years of flying DC-3s and Curtiss Commandos had obviously paid off. We stopped on the runway, about two-thirds along and

The Caribbean was one of the last habitats of the Douglas DC-3 'Gooney Bird', a place with pleasant weather conditions and uncluttered airspace, and in this part of the world the low-tech DC-3 can still make a profit as a passenger liner. Once inside we seated ourselves in the very basic seats and waited for the familiar rumble of starting engines. This would have been very welcome as the temperature inside the cabin was starting to soar already, despite the early hour. Barely in our seat, the flight attendant called us forward into the cockpit area. Naturally we obliged, only to find the cockpit

the islands. Perfectly trimmed a DC-3 can be flown with your fingertips, which is exactly what Hernan did while leaning backwards in his seat in a reassuringly relaxed manner.

All too soon we started our very slow descent towards some specks on the horizon. As we got closer we could make out dozens of small islands in the clear blue turquoise sea.

We passed overhead some luxury yachts, their occupants clearly enjoying the warm sunny weather. As I gazed through the DC-3's square windows I

made a turn to backtrack the runway to our parking stand on the small semi paved ramp. Having disembarked, we collected our bags which were offloaded via the left-hand hatch just aft of the cockpit. The tropical heat was already in force, so it was perhaps time to relax and enjoy a couple of cool Polar beers!

We would like to thank our friends at Lineas Aereas Canedo and the crew of the Sol de America flight for their assistance and some unforgettable moments.

Sol de America brighly colored 1945 DC-3 YV-911C was photographed prior to our morning flight from Caracas International Airport to the tiny holiday resort Island of los Roques, which lies about 128 km North of the Venezuelan coast. At the time Sol de America operated a single DC-3 and a BN-2A Trislander on passenger flights to Los Roques.

Built at the Douglas Oklahoma plant as a C-47B-25-DK Skytrain with (c/n 16013/32761) she was assigned s/n 44-76429 March 1944 and transferred to the RAF as a Dakota IV with tail-code KN387 that same year. After the war she went to the Pakistan Air Force as H-712. Her civil career started in 1954 and she went through several owners including Field Aircraft Services, Societe Aigle Azure, Aer Lingus, SATA Portugal and LANSA Honduras. Caribbean Flights of Venezuela bought her in 1996.

Sol de America ceased all its commercial operation early 2000 and currently YV-911C is stored at Valencia airport Venezuela.

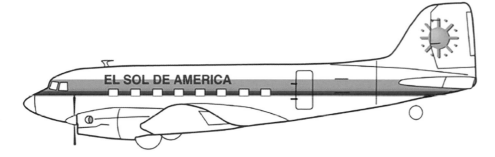

Postcards from Colombia & Venezuela

(South America DC-3 Eldorado)

During the late 1990's it was still possible to observe numerous South American freight companies flying venerable piston engine 'propliners' such as the Douglas DC-3. Several exotic operators were still active in Colombia and Venezuela. It had been my long time desire to visit this continent, and during the month of September 1999 I finally organized a trip 'down south' with a long time aviation friend Andre van Loon. At that time Avensa-Servivensa issued an extensive Air Pass on its route network, and we decided to start our South American adventure from Miami.

We boarded an Avensa Boeing B737-200 classic jet, bound for Caracas. Three hours later, we arrived at the bustling international airport of Caracas–Maiquetia, located on the rugged northern Venezuelan coastline. Expecting a smooth onward transit to Bogota it was here that we encountered the first of a series of travel problems.

Our initial delay turned out to become a 5 ½ hour wait, due to the late arrival and combined technical fault on the incoming B727-200 jet. We spent a total of 7 hours waiting, for a brief one-hour trip to Bogota's El Dorado International Airport.

We arrived very late at night, and the airport was closing down. As soon as we collected our bags, cleared immigration and customs we exited the airport building, only to be jumped by a horde of waiting taxi drivers. We had made an advanced hotel reservation, but our late arrival made me wonder whether we still had a room for the night! I never received a reply to my original fax reservation, and here we were in a strange city late at night, easy prey for the greedy taxi drivers.

Bogota, the nation's capital city has lots off futuristic architecture, a diverse cultural life, splendid churches and impressive museums. It is also a city of beggars, shantytowns, drug dealers and horrendous traffic jams. We did manage to make our way safely to the 'El Presidente' Hotel, which was located in a grim area of the city, but at least we had a place to sleep.

Our plan was to spend only one night in Bogota, and then make our way to Villavicencio. The next morning we inquired if it was safe to make the long taxi ride through the mountains. The friendly hotel taxi driver confirmed that it was still safe to travel, and so we agreed on a price.

Algo más que...
Amanecer llanero

Villavicencio - Meta

We finally arrived at our hotel in Villavicencio and checked in. "Hotel Del Llano" was an international four-star hotel situated on the outskirts of the city. It had all the modern comforts, excellent rooms, large swimming pool, an excellent restaurant and bar. Strangely we were the only international guests staying in the hotel – perhaps it was low season.

Villavicencio was still the DC-3 Capital of the World! However, the glory days of Satena Douglas DC-4s and C-46 Curtiss Commandos were long gone! Only a single C-46 was parked at Villavicencio, but its gorgeous R-2800 radials remained silent and it did not fly. And only two airworthy DC-6 freighters were operational. Thank goodness for the good old DC-3!

A half an hour later we where speeding through the outskirts of Bogota, on our way. Villavicencio is situated about 110 km south east of Bogota, and lies in the foothills of the eastern slopes of the Cordillera Oriental, and is capital of the Meta province. For the most part, the roads were very good, but frequently our driver applied severe braking action in order to avoid two accidents and a mudslide.

The ride took about three hours and we did not encounter any police or military checks. Of course we were well aware of Colombia's political situation, and had wisely made advanced inquiries in regard to overland travel.

Air Colombia DC-3
flight into the jungle
Several months before our trip, I had faxed the Air Colombia office with the intention of requesting permission to photograph their fleet of Douglas DC-3s and DC-6s. Several weeks passed and I had given up all hope of receiving a reply, when out of the blue a fax appeared at my work. It was from Captain Luis Hernan Wilson, president of Air Colombia, stating that we were very welcome and that he would assist us in any way he could. This was very good news. It was essential to have a good contact at the airport, especially when photographing airplanes in Colombia, and it was a long way to travel should our quest have proved fruitless.

The next morning, still shaky from our lengthy taxi ride, we made our way to La Vanquardia Airport! There we got our first glimpse of the numerous hard-working DC-3's. Various companies, such as, Latina, Viarco, Aliansa, Sadelca, Alcom, Ades, Air Colombia and Aerovanquardia were still operating the Douglas DC-3/C-47 workhorse, on daily passenger and cargo flights within the Meta province.

Driving up to the Air Colombia ramp was a bit tricky for the taxi driver. It was a dirt road full of potholes and large stones leading up to the aircraft ramps. We arrived at the gate and rang the bell several times. One of the ground mechanics appeared and opened the gate. Inside we met up with Captain Wilson and introduced ourselves as the two Dutch photographers from Holland. He was very glad

to meet us and in turn he introduced us to one of his DC-3 pilots, Pablo Mauricio Tovar.

After viewing Air Colombia's maintenance facility, Captain Wilson invited us for lunch at the local airport cafeteria. We dined on a typical Colombian meal consisting of rice, baked bananas, yams, cassava and spicy beef.

Air Colombia started back in 1987, when they applied for a general cargo certificate. Two years later they introduced their first Douglas C-47A, HK-3292 (9186), flying out of El Dorado International Airport. During that time Antonio Reyes headed the company, and Air Colombia operated within Colombia, to San Andros, and in addition crossed the border to Panama. Soon after, Air Colombia

received its passenger certificate, which allowed them to carry up to 20 people on the DC-3.

Captain Wilson began flying as a co-pilot on the DC-3, and later progressed to the Douglas DC-6 as captain. He made the news headlines during February 1991, after suffering an incident while fly-

ing Douglas DC-6A(C-118A) HK-1702 (44670). During take-off on the morning of the 10th February, for a routine charter flight from Bogota to Puerto Carreno, with 80 military personnel (including their equipment) on board, he encountered a double engine failure on his initial climb. Captain Wilson was forced to carry out an emergency landing in a grass field that lay before him. After dumping fuel, he managed to land the crippled DC-6 safely. But the aircraft broke up on impact, and immediately caught fire. By some miracle, all 80 passengers and 5 crewmembers walked away alive.

When Transamazonica went out of business Air Colombia took over its hangar and office space, thereby providing the airline with its operational headquarters. Most Air Colombia flights were to remote jungle villages, such as Mitu, Miraflores, Carrenó, Puerto-Inirida, La Padrera and Taraira. For the most distant locations, Air Colombia used its flagship DC-3 HK-3292, due to its long-range fuel tank installation, which is similar to a military

HC-47 variant. A total of 1,600 gallons can be carried, and this enabled it to reach the remote and isolated Indian villages deep in the Colombian jungle.

Two days after our arrival at Villavicencio, Captain Wilson kindly invited us to join him on a regular long-range cargo flight to La Padrera, a small Indian village, located 730 km south east of Villavicencio, very close to the Brazilian border. The only way to get there is by a lengthy voyage on the Japura River or by air, with the latter option being far quicker.

He was one of Air Colombia's check pilots with some 14,000 hours experience on the DC-3. His flying career began with the Colombian Air Force flying all sorts of military jets, including the Boeing 707 airliner.

Once loading of the last cargo had been completed, it was time for us to climb aboard and settle down on one of the side-mounted fold-down seats. The Air Colombia DC-3 mechanic also joined us. He had already run the engines, and checked the systems prior to our arrival.

to "METO power" (maximum except take off power) with setting 42-inch manifold pressure. We continued our slow but steady climb towards our cruising altitude.

Well used to the ritual of DC-3 flying, Captain Wilson busied himself reading the morning paper, while the ground engineer took his morning nap. While talking to Pablo, he told me that only two week's before this flight he had to shut down a bad engine. His DC-3 was losing height quickly, so they decided to throw out the cargo at 2,000 feet! But today, both Pratt and Whitney R-1830 engines were running smoothly….thank God!

Two hours later we began our descent over the thick jungle. We arrived overhead a small 1,700 ft long airstrip located near the Japura River. Captain Joaquin came in overhead and turned for a downwind approach. We swooped

Wednesday the 8th September was an early rise. We had to report to the Air Colombia maintenance ramp at seven in the morning. But when we got there, there was little activity.
A total of 2,600 kg of Coca-Cola cans, eggs, toilet paper, onions, potatoes, baking powder and fresh vegetables were listed on the cargo manifest. Most of the cargo had already been loaded and checked by the airport military, which was standard practice at Villavicencio.

Some 30 minutes later, Captain Wilson and Pablo turned up together with another pilot, Joaquin Hernan Sanclemente.

Also joining us was a family with two children travelling to La Padrera. Captain Joaquin fired up both engines and we taxied down towards the holding point to make our final checks. Two other DC-3s lined up behind us in the early morning departure queue.

As we began our take-off run, both engines roared up to 48-inch of manifold pressure, turning at 2,700 rpm, producing a total of 1,200 brake horse power per engine. With the gear up, we slowly climbed in the warm air and made a gentle right-hand turn towards the southeast. Both engines were throttled back

low over the trees and made a tight turn on to final approach. Some light rain was encountered during the final approach, and as Captain Joaquin eased the throttles fully back we gently touched down on the partially paved landing strip. As we taxied toward the parking spot on the very edge of the strip, the village inhabitants swarmed around the DC-3, all wanting to meet and greet the crew!
As soon as the propellers stopped, the big cargo doors swung open and warm humid tropical air suddenly filled the cabin. A large dilapidated signpost on the local village church read: "Bienvenido La Padrera!"

The local Indian tribes are always happy to watch the arrival of the DC-3. It's a cheerful occasion, and the lifeline DC-3 brings in fresh supplies and visitors from the city. We jumped out only to be greeted by the military who wanted to check our passports! Captain Joaquin explained that we were two photographers from Holland visiting Colombia, especially to see old Douglas DC-3 in action. After a quick inspection we duly received clearance, and joined the crew for the short walk to the local village cafeteria, where we were offered another "spicy" lunch which included snake meat.

Meanwhile, our DC-3 was emptied of its contents, and the return cargo was loaded aboard, comprising 2,900 kg of "Bagre" fish. According to our cheerful co-pilot Pablo, this fish tasted delicious. Its smell certainly filled the DC-3 cabin. We finished our hearty lunch and thanked the old lady for cooking such a wonderful meal. All too soon it was time to head back to Villavicencio with our smelly cargo and further passengers. Our DC-3 struggled to get back in the air from the short airstrip, mainly due to our high weight, but once airborne we climbed to a cruising altitude of 9,000 feet, and it was a leisurely flight back to South America's most famed DC-3 airport.

Servivensa, Parque Nacional Canaima – Venezuela.

Several days later we left Villavicencio on an early morning flight, onboard a modern Aires Dash-8 turboprop, leaving behind one of the world's most adventurous airports thanks to the enduring appeal of the DC-3. The flight to Bogota lasted only 35 minutes, as we crossed the northern Andean mountains at an altitude of about 6,000 metres. What was supposed to be a simple flight connection out of El Dorado Airport and onwards to Caracas, with a connecting flight inland, turned into a two-day travel nightmare. It started with our flight cancellation, an exhausting re-booking, a missed connection, and an unforeseen night stop at Caracas' dirty and smelly seaside town of Macuto!

After a hot and sticky night at a local cockroach hotel, we arrived back at Caracas Int. airport, feeling very exhausted, and boarded our midday flight - an Aeropostal Douglas DC-9-50 jet - to the modest city of Puerto Ordaz (Ciudad Guayana), situated on the South bank of the Orinoco River.

Puerto Ordaz was a newly-developed metropolis area along the Orinoco. Its population already exceeded half a million, and the majority of its inhabitants worked in the iron-ore business. Puerto Ordaz and Ciudad Bolivar are the two gateway cities for the Orinoco delta and the Guayana Highlands aka "Gran Sabana" frontier, which takes up half the country. This area spans about three million hectares, making it the largest national park in the world. With its legendary Tepuis

Mountains, thick jungle, wild rivers, huge rapids and towering waterfalls, this park is also referred to as the "Lost World".

Our main destination was the small tourist lodge camp at Canaima. This camp is located on a beautiful stained lagoon overlooking the spectacular "Rio Carrao" waterfall. There are various excursions from Canaima, which are thoroughly worthwhile. One of them is an expedition into the jungle visiting Indian villages and exploring the world's highest waterfall the "Angel Falls" at 979 metres.

Servivensa operated daily scheduled DC-3 passenger and cargo flights out of Puerto Ordaz to Canaima and Santa Elana de Uairen. Another company, Rutaca, which was normally based out at Ciudad Bolivar, also flew daily DC-3 passenger services to Canaima.

After spending the night at the Hotel Rasil, we returned to the airport for our morning Servivensa DC-3 flight. We checked in for flight VC612 and discovered

that there were only three other passengers on this service, together with a Servivensa stewardess.

At 08.15 our DC-3A YV-147C fired up its radial engines and we began to taxi out to the runway holding point. Then, Captain Miquel Grau and his co-pilot Mauricio Baron, moved the throttles forward for a leisurely take-off run. Once airborne, we made a right-hand turn to the South passing over the city of Puerto Ordaz. Our lightly loaded DC-3 soon settled into a cruising altitude of 8,500 feet. Twenty minutes later we then flew over Venezuela's biggest inland lake, the "Embalsa de Guri". The lovely Servivensa stewardess was dressed in a light brown khaki uniform reminiscent of the 1950s art-deco period. The flight lasted about an hour and you would have thought she would have looked after her three thirsty passengers. But she never once came around with any coffee or drinks – perhaps she did not want to soil her smart uniform!

About forty minutes into the flight, the landscape beneath the DC-3's silver wings changed dramatically with the appearance of the flat top Tepuis Mountains in between endless jungle.

Captain Grau pointed out that Canaima airstrip was shortly coming up on the horizon. And indeed, we soon arrived overhead the lagoon, with its thundering waterfalls. We made a downwind approach, turned into the wind for a smooth landing, and taxied onto the small parking ramp, which was made from red gravel. Once the engines had been shut down, we disembarked through the aft door and stumbled into a party of European and Japanese tourists. They were waiting to go on a local sight-seeing flight over the world's highest falls, the "Salto Angel", in a DC-3!

Servivensa DC-3s were all painted in a smart blue and white 1950's Pan American-style colour scheme. We watched as Servivensa DC-3 YV-609C fired up

her Pratt & Whitney R1830s radial engines and took off for another sight seeing flight. Afterwards it was time to check in at our modest room in the adjacent village behind the airstrip arranged by Canaima Tours. For the next two days we enjoyed the primitive and relaxed atmosphere in and around the camp, while watching the movement of the local Antonov An-2s and Douglas DC-3s.

It was time to leave, and continue on to the next stage of our South American adventure. To my surprise we were the only passengers travelling back to Puerto Ordaz, accompanied by a second DC-3 crew. Veteran DC-3 captain Miquel Grau was also on board checking the DC-3 flight operation for Servivensa. It felt as if this was a private flight for the two Dutch international travelers.

DC-3 YV-610C awaited us on the dusty gravel strip. We settled in to our seats, and the DC-3 fired up her radial engines.

During a relaxed take-off I viewed the jungle, sinking lower and lower under the massive wing of the DC-3. We turned to the north and I had my last view of impressive and exotic Canaima.

I had seated myself in the forward cabin for a grand view of the left-hand engine together with the spectacular scenery that unfolded below. After take-off I noticed a light smoke trail emanating from the engine, but thought nothing of it! About 15 minutes into the flight I noticed Captain Grau entering the cockpit. Soon after a light shudder went through the cabin, and then I noticed that the port engine had been shut down and the gleaming propeller blades had been feathered! Not wanting to risk the remainder of the flight - we still had a good 45 minutes of flying to reach our destination - the crew prudently made a gradual 180 degrees right-hand turn for a straight-in approach into Canaima. We made a steep approach and landed safely on Canaima's 2,500-metre long runway. Taxying onto the parallel gravel taxiway with the use of the starboard engine, we tried to reach the main parking area, but to no avail! YV-610C would not co-operate and the crew called it a day and shut down the remaining engine.
The Servivensa crew had done an excellent job getting the DC-3 back to Canaima. After our return the captain arranged

for a substitute DC-3 (YV-147C) to fly us back to Puerto Ordaz that same afternoon to conclude another chapter in our South American odyssey.

Aeroejecutivos DC-3 to Los Roques Archipelago
Our Airpass centred around Caracas International Airport, and we needed to get back there for our next adventure. We decided to travel by bus from Puerto Ordaz to Ciudad Bolivar, where we then boarded a sleek Servivensa B727-200 jet back to Caracas. Inside the domestic terminal, Venezuela's third DC-3 operator operated a small ticket counter for its daily flights to Los Roques. Aeroejecutivos (AE) operated several DC-3s and a single Convair 440 from its Caracas base. During our visit it was possible to buy a return ticket for 80 dollars to the tiny island paradise of Los Roques, situated about 130km to the north.

This island chain off the Venezuelan coast has much to offer, and consists of 42 islands surrounding a 400-km lagoon. Most Islands are uninhabited. The largest of them "El Gran Roque" consists mainly of a small fishing town, beaches and a short airstrip. There were already many tourists visiting to enjoy the marine life. It is a paradise for scuba divers and sport fishermen.

AEROEJECUTIVOS

HERMANN ZINGG
President

N.W. 64 AVE. PERIMETER RD.
BLDG. 1008 • MIAMI, FL 33122
(305) 871-2099 • FAX: (305) 871-2095

BFM LA CARLOTA
ZONA COMERCIAL AEROCLUB
CARACAS LOCAL NO. 4
CARACAS, VENEZUELA
TEL. TEL. 91 7942 • FAX: 91 6215

We checked in at the AE ticket counter, requesting permission to photograph their brightly coloured DC-3 and Convair 440 fleet. One of the attractive dark haired Aeroejecutivos representatives helped us out and arranged a ramp tour. Usually the AE DC-3 was parked on the business ramp, in between the executive jets, in preparation for the morning flight to Los Roques.

Together with flight attendant Victoria Pereira we obtained permission to walk up to the immaculate Douglas DC-3A registered "YV-440C" (2201) basking in the bright morning sunlight. This particular DC-3 G-102 was originally delivered new on 29th of February 1940 (with a right-hand passenger door), fitted with Wright Cyclone radial engines, to American Air Lines. Nine years later she was converted to a DC-3A for Trans Texas Airways. This veteran DC-3 had retained her right-hand passenger door and subsequently flew for the famous Provincetown Boston Airlines (PBA) as N31PB. Today, she stood proudly amongst the sleek executive jets awaiting another group of passengers.

Once I got out my camera, one of the ever-vigilant security guards shouted, "No Photography!" thus ended our photo shoot. Very disappointed we returned to the terminal. Victoria then approached me and said that she would talk to the Captain. "Give me one moment please." Later she returned with a very youthful looking Captain Luis Piccardo (aged just 29) and his co-pilot, Carlos Sarzalejo. "There is no problem," he said, "why don't you come with us to Los Roques and photograph the DC-3 there." That was a very generous offer which we gladly accepted. "We will arrange for some tickets for you and your friend." Thirty minutes later we joined the other passengers aboard YV-440C, which was also known as "Caballo Viego". Interestingly, the cabin featured white walls, oak wood paneling and camouflaged cabin seats! Engine start-up and taxi out was pretty much routine. We took off from runway 09 and made a gentle left-hand turn towards the open ocean for a relaxed one-hour flight to the spectacular Gran Roque Archipelago.

We landed smoothly on the 1,000 metre-long airstrip and taxied to the small gravel ramp in front of the open air terminal, which boasts a big sign declaring 'Bienvenidos a Los Roques'. After disembarking from this trusty old DC-3, we soaked up the cool tropical breeze and enjoyed the picturesque scene. This tiny Island has no cars and is alcohol free.

We did not stay on the Island and hanged around the DC-3 for some photo options, which included the friendly crew Captain Luis Piccardo, co-pilot Carlos Sarzalejo and cabin FA Victoria Pereira. After the return passenger boarded the aircraft it was time to head back to Caracas Maiquetia - Simon Bolivar airport. We made a grace-full takeoff and made a right-hand turn over deep blue Caribbean waters. After a couple of minutes the roar off the Pratt & Whitney R1830s radial engines eased a bit as we entered our cruise back to Caracas.

Venezuelan DC-3 freighter YV-1854 (c/n 6135) was photographed at Opa Locka Airport FL during an annual aviation trip back in June 2010. I have photographed this a/c before when she with **Aeroejecutivos** as YV-500C based at Caracas Maiquetía "Simón Bolívar" International Airport Venezuela September 1999.

She rolled of the production line for the USAAF with serial number 41-19492 in January 1943. She was transferred to the 10th AF India that same year and later to India China Wing ATC.

During her civil career, which started in 1950 she flew with Hindustan Aircraft Co as VT-DGU and later with Ariana Afghan Airlines as YA-AAC.

In 1973 she arrived in Canada and flew with Lambair Ltd and Perimeter Airlines as CF-DBJ/C-FDBJ.
While operating with Lambair she suffered a painful incident at Tadoule Lake Manitoba, during the winter when she fell through the ice on her belly. Luckily she was recovered with no major damage.

In 1988 she was sold to Aeroejecutivos SA as YV-500C and moved to South America. Early 2007 she was noted at Pompano Beach Airport FL with a new registration YV1854. She remained unsold up until 2013, when she was broken up for parts.

Top: Museo Aerospecial Colombiano, Catam, Bogota

This was the author's second visit to Vil-lavicencio Vanguardia Airport, the "last stronghold of commercial Douglas DC-3's" in South America. During the late 1970s early 1980s Vanguardia was a piston-engine propliner 'El Dorado' with numerous local operators flying cargo and passenger DC-3s to the far-flung native Indian villages scattered throughout the jungle of Colombia. Together with a friend we flew to Bogota International airport and traveled onwards to Villavicencio, over the Andean cordillera onboard a local bus service, in order to inspect the latest situation of the current DC-3 operators April 20215.

air colombia s.a.s

Campo Elías Cendales
Despachador

Tels.: 664 8076 - 664 8129
Fax: 664 85 81
Aeropuerto Vanguardia   Cels.: 311 8124746 - 311 2634046
Villavicencio   E-mail: aircolombiasas@gmail.com

Air Colombia Ltda highly polished HK-3292 C-47A-80-DL (c/n 19661) was photographed at her home base at La Vanguardia Airport, Villavicencio Colombia, she was delivered to the USAAF back in February 1944 with

serial number 43-15195. Served in the US as N87645 with Aero American Corp and later in Canada as C-GABG with Air Brazeau Inc It entered Colombian register in 1988 with Air Colombia/Aeroville as HK-3292X and mainly used as a mixed passenger-freighter. Sadly the situation in Colombia has worsened and Air Colombia was forced to close its doors. All DC-3 were offered for sale March 2019.

Air Colombia Ltda sistership HK-1175 (c/n 20432) was originally built as a C-47A-90-DL back in May 1944 with serial number 43-15966. After military service it operated with RW Duff (N5592A) and Miami Airlines Inc as N3935C back in the late 1950s and in addition served in Canada with Montreal Air Services Ltd (1956) as CF-DME. She operated in Colombia with Lineas Aereas La Urraca from 1964 to 1975 and later flew with Transamazonica.

***Bottom:* Aerolineas Andinas** - Aliansa SA HK-5016 (c/n 14101/25546) was originally built as a piston C-47B-1-DK, July 1944 with serial number 43-48285 and was drafted into the French Airforce with serial number 348285. Later it became F-BTDE with SASMA, "43-48285" Yugoslav Airforce, YU-ABV with OCZS and N8071Y with Atlas Aircraft Corp. Sold to SAAF in 1981 with tail code "6880". During 1995 it was converted to a C-47TP and sold to Dodson International Parts Inc as ZS-OJL/ZS-OJM at Wonderboom Airport South Africa. Returned to

Ottawa Kansas in the US for onwards sale to Aliansa late 2013, it remains the only civil turbine DC-3 in Colombia. She was involved in a nasty landing incident at San Felipe Colombia when she veered off the runway, which resulted in a RH main gear collapse and damage to both engine and props. (08 April 2022).

Top, right: HK-2820 (c/n 20171) photographed at the company base at La Vanguardia airport, Villavicencio was painted in semi black and white D-day colours. Built in 1944

as a C-47A by Douglas Longbeach factory she was delivered to the USAAF with serial number 43-15705. During the 1950s she flew with Northwest Airlines as N79055 and then went to Ozark as N151D. Acc to the records she was impounded at Bogota (1976) and several years later received a Colombian registration HK-2820 and joined Aliansa. Sadly on the 8th of July 2021 she crashed on a training flight North of Villavicencio killing all three crewmembers, Captain Jhon Acero, Captain Carlos Cortes & technician Carlos Olaya.

Aerolineas Del Llano SAS, ALLAS operated two DC-3 freighters from their home base La Vanquardia – Villavicencio airport during our visit in April 2015. Its fleet consist of HK-1315 (c/n 4307) delivered to the USAAF with serial number 41-7808 back in April 1942. Before operating in Colombia she used to fly in Brazil with REAL and Varig as PP-ANG. It first arrived in Colombia back in September 1966 with Avianca. It went through numerous operators such as Lineas Aereas la Urraca, Aerolineas Santa Ltda, Taxi Aereo El Venado, Transamazonica before ending with her current owner ALLAS.

ALLAS 2nd ship HK-3215 (c/n 14666/26111) is an interesting aircraft. It carries a data plate showing it was built at the Oklahoma plant as a DC-3C per September 1944. According to the Air Britain DC-3 bible she was built as a different model: a C-47B-5-DK for the USAAF with serial number 43-48850 with the same delivery date September 1944. She was transferred to the RAF Montreal as JK938. She operated with Air Atlantique Ltd as G-ANTC and as N4261P with Sky freight Airlines Inc based at Haileah Florida (1983). She arrived in Colombia early 1986 with LACOL as HK-3215X.

Aerolineas Llaneras Ltd - Arall HK-2663 (c/n 12352) C-47A-5-DK rolled of the production line in January 1944 with serial number 42-92540 and served with the RAF Montreal as KG345 later with CAF as '12352'. After the war she went to Canada serving Ontario Central Airlines as CF-XUS.

Sold to Colombia in October 1981 with registration HK-2664X to Lineas Aereas Eldorado Ltda. Later on it served with ADES and Aerovanquardia. Regrettably in April 2016 she had to make an emergency after take-off from Puerto Gaitan due to engine problems. All three crew members were uninjured and evacuated while the aircraft was destroyed by fire.

Aerolinea Llaneras Ltda.

Sergio Cruz Zapata Parales

Aeropuerto Vanguardia Hangar Norte - V/cio.
✆ 6648446 - 6648472 Cel.: 310 8658927

Sociedad Aerea Del Caqueta - Sadelca Ltd active Skytruck HK-2494 (c/n 16357/33105) was built as a C-47B-30-DK but delivered as a TC-47B to the US Navy as a R4D-7 BuNo 99826. Her military career ended in 1971 and she was sold off to the University of Texas in Austin TX as N87611. Sold to Colombia in 1980 to SAEP Ltda and based at Bogota International airport. Early in 2019 she was sold to LASER Aereo Colombia. Sadly on the 9th March that same year she was involved in a tragic landings accident en-route to Villavicencio Airport, killing all 14 crew and passengers. Due to this accident the Colombian CAA became stringent on the DC-3 operations in Colombia.

RUTAS
Pto. Inirida
San Jose del Guaviare
Mitu
Caruru
Miraflores
Barrancominas
Mapiripan
Taraira

Aerocharter del Oriente

Top: Sadelca R4D-6 HK-1149 (c/n 26593) was photographed during maintenance at the company base at Villavicencio. Constructed by Douglas Oklahoma plant she was delivered with USAF serial number 43-49332 and converted soon after to a R4D-6 for the USN with Bu Number 50795.

AIRSHOWS EUROPE

During September 2007 **Lufthansa Technik** AG and Hamburg Airport celebrated their annual Aviation "Airport Days" event, with live music and a large gathering of vintage airplanes, which included seven DC-3s, a rare Douglas DC-2 and the world's only airworthy Lusinov LI-2.

On the left page from left to right: DC3 Association F-BBBE, Air Atlantique G-AMPY, Aviodrome DC-2 NC39165/PH-AJU and Flygande Veteraner SE-CFP.

On the Right page from right to left: Vallentuna Aviators 9Q-CUK, Dakota Norway LN-WND and Vennerne Club OY-BPB

The **Aviodrome Museum** located, at Lely-stad Airport Holland, hosted its first ever Douglas DC-3 "Dakota Fly-In" during the weekend of 27/28 of May 2006. Organized by the Aviodrome itself, it invited all the operational European DC-3s to spend the weekend in Holland.

The Aviodrome static DC-3 was open for visitors, while its DC-2 was inside for maintenance. Due to bad weather on Saturday morning some participants were unable to make it. DDA Classic Airlines two DC-3s were detained for various reasons and the French DC-3 returned back to France when weather at Lelystad was below landing minimums on the Friday. Highlight of the event was the arrival of the world's only operational Lisunov Li-2 from Hungary. Other DC-3s arrived from UK, Norway, Sweden, Finland and Denmark.

Daks over Normandy - 70th Anniversary of D-Day Normandy - Cherbourg 04-06 June 2014

Amazing line-up at the 2014 Daks over Normandy, which celebrated the "70th Anniversary of D-Day" at Cherbourg Airport Normandy. From Left to right: The Danish Dakota Friends (DC-3 Vennerne) OY-BPB, Dakota Norway Foundation C-53D LN-WND and Association France DC-3 F-AZTE.

The "70th Anniversary of D-Day", was celebrated with a gathering of several DC-3s and C-47s at Cherbourg-Mauperus Airport, Normandy. Under the banner "Daks over Normandy" 10 aircraft made it to Cherbourg, which lasted from 4-6 June 2014. Two C-47 Skytrains N345AB aka "Whiskey 7" and N74589 "Placid Lassie" (ex-Union Jack Dak) flew over from the USA. Also attending the event was the only operational Lusinov LI-2 from the Goldtimer Foundation HA-LIX.

The 'Round Canopy Parachute Team' (RCPT) was also involved and they invited all resident UK and European Dakota's to Normandy to re-enact the C-47 Allied paratrooper missions that took place during the early

morning hours of 5/6 June 1944. A total of 13.000 plus paratroopers then jumped out of a grand total of 821 Douglas C-47s behind enemy lines.

On the 4th June 2014 all aircraft's assembled at at Lee-On-Solent Airport, UK and made the historic cross channel flight and flew onwards to Carentan Normandy for the historic drop. Due to the high winds, the drop was cancelled and the armada flew safely back Cherbourg-Mauperus Airport.

The 1940 immaculate Breitling DC-3A HB-IRJ was the only aircraft not capable of dropping paratroopers, but instead it was used for memorial flights over the Normandy beaches.

One of the main events organized by the Daks over Normandy team was an evening photo-shoot for the international press and photographers. Six aircraft were on display, nearby the Duxford Imperial War Museum main ramp and were lighted up with flood lights. The evening went well with no wind and clear night skies. Star attraction were several US based C-47s such N24320 "Miss Montana", N45366 "D-Day Doll", N47SJ "Biscuit Bomber" and N47TB "That's all Brother".

Daks over Normandy - 75 years D-Day

It had been several years in the planning and full scale preparations were made for the 75th Anniversary of "Operation Overlord", the D-Day landings on the 6th June 2019. It was announced as a "Grand Arial Spectacle" in the skies above Duxford, England and Caen, Normandy France, a sight not seen since WW2. The US "Confederate Air Force" (CAF), The "D-Day Squadron/Tunison Foundation" and from Europe the "Daks over Normandy"(DON) all joined forces to make this a very special event.

The main aim was to assemble a fleet of 25-30 Douglas DC-3s/C-47s, flying over the

skies of England and France and to recreate military style jumps over the historic Drop Zone (DZ) at Sannerville (Normandy) in honor of the troops who bravely did so 75 year ago.

The D-Day Squadron fleet, the US contingent of 15 DC-3s and C-47s, also referred to as the "Mighty Fifteen", successfully made it across the North Atlantic. There route was also used in WW2 and known as the Blue Spruce Route, and routed via Waterburg- Oxford airport (Connecticut), Goose Bay (Newfoundland, Canada), Narsarsuaq (Southern Greenland), Reykjavik (Iceland) and Prestwick (Scotland). All other DC-3s arrived from France, Hungary, Swiss, Scandinavia, Denmark, Finland Holland and England.

The official event took place at the Imperial War Museum at Duxford, Cambridgeshire and lasted from 2 to 5 June and from 5 to 9 June at Caen Carpiquet Airport Normandy France. The official Cross channel flight took place on the 5th June when the DC-3/C47 armada flew from Duxford via Colchester, Southend-on-Sea, Maidstone and Eastbour-

ne, heading out over The Channel (North Sea). They passed Le Havre and then went onwards to the historic UK Drop Zone at Ranville and Sannerville.

US Dakota Fleet:
N47E 'Miss Virginia'
N47SJ 'Betsy's Biscuit Bomber'
N47TB 'That's all Brother'
N877MG 'Historic Flight Foundation'
N18121 'Blue Skies Air'
N45366 'D-Day Doll'
N8336C 'The Spirit of Benovia'
N24320 'Miss Montana'
N103NA 'Flabob Express'
N62CC 'Virginia Ann'
N74589 'Placid Lassie'
N150D 101st Airborne Tribute
N341A Golden Age Air tours
N25641 Legend Airways
N33611 Clipper Tabitha May

European Dakota Fleet:
F-AZOX Chalair
SE-CFP 'Daisy'
LN-WND Dakota Norway
OH-LCH Aeroveteran OY
OY-BPB Danish Dakota Friends
N431HM Classic Formation Team Swiss
HA-LIX Goldtimer Foundation
PH-PBA DDA Classic Airlines
N147DC Aces High
N473DC Aero legend
G-ANAF Aero Legend

AIRSHOWS USA

Warbirds & Legends Gathering (WLG) at Forbes Field – Topeka, Kansas (August 2013)

The "Warbirds & Legends Gathering" (WLG) "theme", was a reunion with people that were interested in aircraft of the 1930's and 40's, as well as their current owners and pilots that operate these aircraft today. Plus anyone that appreciated the period in American History around the time of World War II.

Warbirds & Legends Gathering was a rare opportunity to see these aircraft in action, and hear what they sound like. The WLG organizers, Dan Gryder, Brooks Pettit, Scott Glover and Robert Rice, invented this new concept of, 'An airshow that's not an airshow", which would include normal aircraft operations.

During all three airshow days the organizers catered for lunch and dinner for all media, photographers, crews and slot-holders. The food and drinks was excellent and a big surprise was the portable beer tap which was stationed in a corner of the hangar! This turned out to a big hit.

This event was certainly a bold idea, and I would like to thank the organizers; Dan Gryder, Brooks Pettit, Scott Glover and Ro-

bert Rice for their help and support. I had a great time meeting up with new friends and took some great pictures.

Warbird and Legends Gathering attracted numerous DC-3s & C-47s from around the country, in total nine aircraft made it. This line up shows N87745 "Southern Cross", N2805J "Spooky", N47HL "Bluebonnet Belle", N5106X "Sky King" and N583V Airborne Imaging.

During this event the WLG team organized several air-to-air sessions for the local and international press. One of the camera ships used was the historical Mid America Flight Museum immaculate Lockheed C-60A Lodestar N1940S with the owner Scott Glover flying it. In addition the Herpa Wings 1938 Douglas DC-3 N143D flown by Dan Gryder was also used.

Participant aircraft were: N47E 'Miss Virginia', N47HL 'Bluebonnet Belle', N143D 'Herpa Wings', N346AB 'Spirit of Hondo', N583V 'Airborne Imaging', N877MG 'Panam', N2805J 'Spooky', N5106X 'Sky King' and N87745 'Southern Cross'.

Top: Herpa Wings N 143 D

Centre: N5106X 'Sky King'

Right: N28053 'Spooky'

'Wings over Dallas' Commemorative (CAF) airshow (26-28 October 2018) at Dallas Executive Airport Texas

The 'Wings over Dallas' WW2 airshow, was slated to be the opportunity to view and watch authentic, World War 2 aircraft in action. Several CAF units brought there aircraft to the show, such as the heavy bombers group, which included the Boeing B-29, B-24 and B17 plus an array of fighters and twin bomber & transport aircraft's. The central theme: included the CAF American Airpower Demonstration, American Armor & USO show, and the Aero-shell Aerobatic team with the North American AT-6 'Texan'. The show concluded with a performance of the "Ladies for Liberty" performing patriotic songs in the style of the Andrew Sisters. Throughout the 3-day event numerous aircraft were available for rides and that proved to be very popular with the public including yours truly.

For me the most important theme was the 'Tribute to D-Day and Normandy' with pathfinder pilot Dave Hamilton giving comments and a quartet of Douglas Skytrains and Skytrooper in action. In addition the Frederick (OK) based WW2 Airborne Demonstration Team (ADT) & The WW2 Liberty Jump Team where in attendance to pay tribute to the men and women that jumped on the eve of D-Day at Normandy June 6th 1944. Both teams jumped from all four aircraft during the afternoon show in traditional WW2 uniforms and gear.

- N47TD (ex N88874) Douglas C-47 Skytrain (c/n 12693) CAF "That's All Brother"
- N45366 Douglas C-53 Skytrooper (c/n 11757) CAF Riverside California "D-Day Doll"

- N87745 Douglas C-49J (c/n 6315) Greatest Generation Aircraft
- N151ZE Douglas R4D-6 (c/n 26408) "Ready For Duty", Commemorative AF Dallas-Forth Worth Wing.

EVENT SCHEDULE

May 19th-21st, 2017
Riverside, California

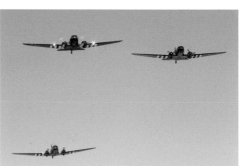

The little airport that time forgot...

2017 – Los Angeles road trip Flabob Fly In

During May 2017, Flabob Airport, near Riverside, CA hosted its first large scale Douglas DC-3 "Dakota Fly-In". Under the command of local business man Jon Goldenbaum Ceo of Poly Fiber Aircraft Coating and the Flabob Airport Authority; all airworthy US DC-3s were invited.

Supported by local businesses, the Flabob Airport Café, Tom Wathen Center, EAA Chapter One and DC-3 and Beech 18 specialist "Garsa Air Inc" this event developed a 3-day extravaganza.
The programme included Breakfast at the Flabob Café, Dakota Film Festival, DC-3 maintenance Seminar, Beer Garden and Food vendors, Aircrafts display and tours, plus DC-3 rides!

The local based DC-3, the "Flabob Express" N103NA, was joined by the beautifull privately owned Benovia C-53D N8336C, which was fitted out with a luxury VIP interior. Two WW2 C-47 Skytrains painted in D-day colours, the Lyon Air Museum C-47B N791HH and the Tunison Foundation C-47A N74589 shared the ramp side by side. Another rare WW2 veteran showed up, this was the world's only operational and privately owned HIPER DC-3-R2000 N62CC. A second rare bird made an appearance, which was

the San Francisco based "Golden Age Air Tours" C-41A N341A. She was one of the highlights of the show. She was giving local rides and many passengers enjoyed her air-conditioned VIP executive cabin!

In total 10 aircrafts from California and across the US showed up. It was a great place to meet with DC-3 owners and crew's and like-minded aviation friends and fellow DC-3 enthusiast.

THE LEGEND AT NIGHT

Top:
Sunrise Schiphol Int. Airport DDA DC-3 PH-PBA, November 2014

Left & Below:
Sunset flight C-41A N341A Riverside CA 2017

Right & Middle:
Madurodam Museum DC-3 PH-APM, February 2022

Bottom: Sunset 'Lady Luck' N408D, Ottawa IL August 2013

Cockpit view DDA Classic Airlines DC-3 PH-PBA.
Flight between Amsterdam and Maastricht 28 august 2021

FLYING DC-3S IN ALL VARIANTS

14 JANUARY 2020

Some 170+ DC-3s in all variants are flying on a regular basis. Their registrations are given below. Registrations with the mark (?) are not fully confirmed and (*) were involved in a recent accident/crash.

Disclaimer (October 2022): By no means this list pretends to give the full picture of all active/flying DC-3s across the globe. It has been compiled by various persons and international sources. The DC-3 community is very much alive with frequent movements/updates and it would be hard to pin down an exact number.

Enlist to fly with the U.S. Navy

Depend on *DOUGLAS*   First in Aviation

- **Australia:** VH ABR, VH AES, VH-AGU, VH-EAE, VH-EAF, VH-MMA, VH OVM, VH TMQ
- **Bolivia.** CP-2421(?)
- **Canada:** C-FBKB, C-FDTD, C-FGCX, C-FKGL, C-FLFR, C-FMKB, C-FTGI, C-FTGX, C-FVNS, C-FWUI, C-GAWI, C-GDAK, C-GEAI, C-GEAJ, C-GGSU, C-GHGF, C-GJKB, C-GKKB, C-GOOU, C-GPNR, C-GRSB, C-GVKD, C-GWZS(?)
- **China:** N41CQ
- **Colombia:** FAC1654, FAC1658, FAC1667, FAC1681, FAC1683, FAC1686, HK-1149, HK-1315, HK-2006 (*), HK-3215, HK-3286, HK-5016(*), PNC-0213, PNC-0256, PNC-0257(*), PNC-0258
- **El Salvador:** YS-347E
- **Finland:** OH-LCH
- **France:** F-AZOX, F-AZTE
- **Germany:** ZS-NTE
- **Hungary:** HA-LIX (Lisunov Li-2, license built DC-3)
- **Iceland:** TF-NPK
- **India:** VP905
- **Mauritania:** 5T-MAH
- **Netherlands:** PH-PBA
- **New Zealand:** ZK-AWP, ZK-DAK, ZK-JGB
- **Norway:** LN-WND
- **Russia:** RA-05738(?), RA-2059G, RA-2944G(?)
- **South Africa:** SAAF6814, SAAF6825(?), SAAF6828(?), SAAF6839, SAAF6852, SAAF6885, SAAF6887, ZS-ASN, ZS-BXF, ZS-CRV, ZS-DIW(?)
- **Sweden:** SE-CFP
- **Switzerland:** N150D, N431HM
- **Thailand:** RTAF46151, RTAF46153, RTAF46154, RTAF46156, RTAF46157, RTAF46158, RTAF46159

- **Turkey:** N3291
- **United Kingdom:** G-ANAF, N147DC, N473DC, ZA947, G-AMPY
- **USA:** N103NA, N115U, N131PR, N138FS, N143D, N144WC, N146RD, N147AZ, N151ZE, N15MA, N15SJ, N17334, N173RD, N18121, N198RD, N1XP, N200MF, N231GB, N24320, N25641, N26MA, N271SE, N272R, N272ZZ, N2805J, N28AA, N28TN, N3006, N300BF, N30TN, N3239T, N33611, N33632, N33644, N33VW, N341A, N345AB, N353MM, N400MF, N4089T, N43XX, N44587, N45366, N467SP, N472AF, N47E, N47SJ, N47TB, N500MF, N5106X, N534BE, N583V, N59314, N60154, N61981, N62CC, N64766, N64767, N64784, N705GB, N728G, N734H, N737H, N74589, N751A, N763A, N791HH, N7AP, N8336C, N834M, N836M, N86584, N8704, N87745, N877MG, N882TP, N92578, N932H, N982Z, N99FS, N34DF and N983DC

- Basler Conversion #68 N1350A (Aug 2022)
- Basler Conversion #69 N941AT (Aug 2022)
- Basler Conversion #70 N856RB ex CF-YQG (Aug 2022)
- Basler Conversion#71 N700CA (Aug 2022)

CANDIDATES FOR POSSIBLE RETURN
- N129H 'Mr. Douglas' skydiving
- N308SF 'Night Fright' 42-100521 79TCS
- N346AB CAF Highland Lakes Squadron, 'Texas Zephyr'
- N514AC D-Day Wings
- PH-DDZ Aviodrome Museum
- NC39165/PH-AJU DC-2 Aviodrome Museum

UPDATES
Check for frequent updates:
- Facebook Group "The Douglas DC-3 Appreciation Society" https://www.facebook.com/groups/DouglasDC3AppreciationSociety/
- WIX-thread "Worldwide Numbers of Warbirds Flying by Type" http://www.warbirdinformationexchange.org/phpBB3/viewforum.php?f=3
- Ralph M. Pettersen's Propliner Information Exchange http://www.proplinerinfoexchange.com/1-dc-3_news.htm

SPECIAL THANKS TO ☺
- John Terrell (WIX)
- Andre van Loon
- Edoardo Poma
- Ralph M Pettersen
- Pete van der Spek
- Ray Watts
- Steve Ozel &
- DDA Magazine Logboek (Jan Willem de Wijn, Paul van den Berg, Paul van der Horst)

------ *Original list: Coert Munk, 14 Jan 2020 - Updated by Michael S Prophet October 2022* ------

INDEX

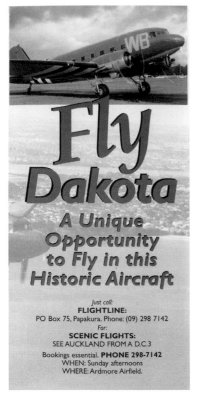

O

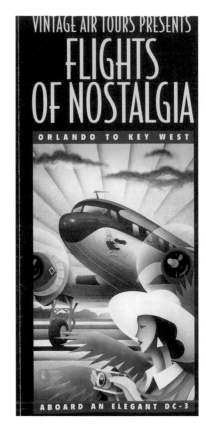

P

R

S

Douglas DC-3, SE-CFP, "Daisy"

Opereras av
Flygande Veteraner

Dakota Norway

**Let the time
fly by**

www.dakotanorway.no

DC-3A

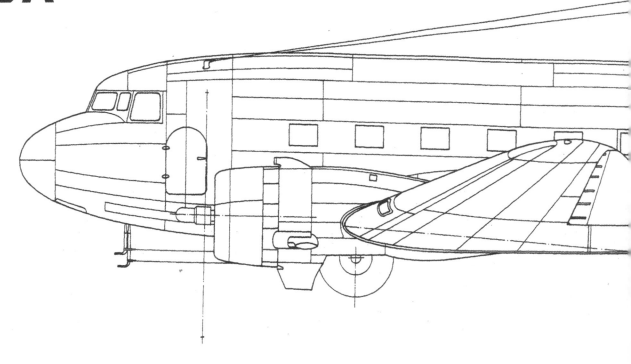

C-47 A/B

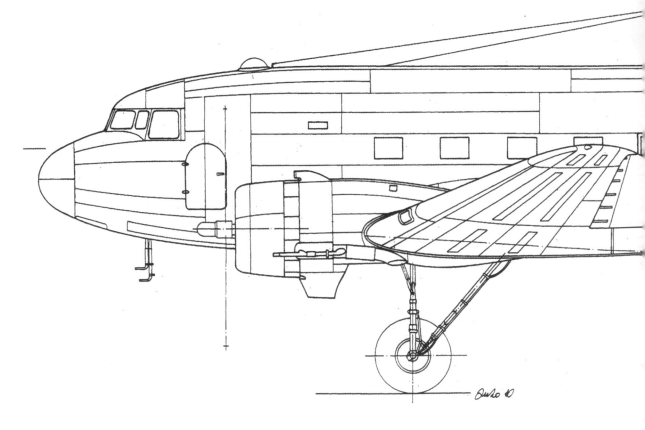